P9-DII-515

CONTENTS

PUBLISHER'S NOTE

Anatomy is the study of the structures of the body. Not surprisingly, an accurate, scientific anatomy did not develop until relatively late in history, with Andreas Vesalius (1514–1564), and even then only against great opposition and religious persecution. Vesalius himself was sentenced to death by a court of the Inquisition; he had the good fortune of having his sentence commuted into a journey to the Holy Sepulcher, but the bad fortune of dying in a shipwreck on his return. Prior to Vesalius there had been several prominent people (Leonardo da Vinci among them) who had contributed to advancing our knowledge of the human body, but by and large anatomy suffered from a dogmatic allegiance to the ancient teachings of Galen (c.130–c.200). Galen's work, to a twentieth-century observer, appears to have suffered from its author's dedication to certain abstract metaphysical principles that, among other things, led Galen to describe parts of the body that did not exist. Vesalius's careful empirical research and rigorous dissection of cadavers served to repudiate Galen and prepared the way for the advent of scientific physiology with William Harvey (1578–1657), who made the very important discovery of the circulatory system and foresaw many of the microscope's anatomical insights.

With the work of Charles Darwin (1809–1882) biology gained a comprehensive framework, and anatomy benefitted greatly, for Darwin's discovery of the mechanism of natural selection grew in part out of his study of anatomical relationships among plants and animals. There is an essential unity and common ancestry for all organisms, Darwin showed, and the evolution of a species involves progressive anatomical changes. You may glean some of human evolutionary history when you read and color Plate 35 and contemplate our obsolete appendix.

The twentieth century has seen a new emphasis in anatomy on living rather than dead animals; X rays now perform some of the investigations formerly accomplished only with the dissector's scalpel. Indeed the very term *anatomy,* which derives from the Greek word meaning "to dissect," is now somewhat out of date. The new emphasis on living tissues is very important because, besides the fact that it is the living and not the dead we care most about anyway, some organs change size, shape, and location immediately after death. To the anatomist interested in the living man, a cadaver is but a shadow of its former self.

Since the body's structures cannot be understood independently of what they do, in the drawings, text, and coloring guides of the *Human Anatomy Coloring Book,* Margaret Matt and Joe Ziemian have occasionally forayed into physiology, the study of the body's functions. The forty-three plates, all of which have been reviewed by a specialist, are divided into the body's systems; hence a reader and colorist will be able to learn about the body's functions one at a time, while working through the entire book will give the fundamental groundwork for more sophisticated studies of human anatomy. The drawings are designed to provide true anatomical detail at the same time that they render organs and other structures clearly, with an eye toward the drawings' usefulness for instruction and coloration. The coloring keys provided are not intended to be naturalistic; rather, wherever possible, they have been developed with the functional relationships of the body in mind.

There is no better time to learn about human anatomy than now, when we are constantly bombarded by new developments in the burgeoning health field, and there is no better way to learn than with the *Human Anatomy Coloring Book,* a useful, accurate introduction to the human body for children and adults, and an enjoyable coloring book as well.

SYSTEMS OF THE HUMAN BODY

The human body is marvelously complex, and the greatest wonder is, complex as it is, how well it works most of the time. For purposes of study, we can divide the body into systems, though we should not forget that each system is itself highly complex and the dividing line between systems may not be distinct. All of the systems have specialized functions, but they are also closely related to one another; indeed their successful interaction is absolutely necessary for our survival.

The *skeletal system* refers chiefly to the bones that support and protect the body. All the muscles that push and pull the skeleton make up the *muscular system*. The *circulatory system* consists of the heart and the tubes—arteries and veins—that transport blood. We breathe with our *respiratory system*, which supplies oxygen to the body's tissues and removes some wastes. The *nervous system*, whose primary components are the brain and the spinal cord, is our "master control," regulating all of our internal functions and providing us with information about the environment. We process food and eliminate some wastes with the *digestive system*. The *urinary system* is responsible for the elimination of most of the body's liquid chemical wastes. The *reproductive system* consists of those organs that characterize the sexes and enables us to conceive, bear, and give birth to offspring. The secretion of hormones, which regulate the body's functions chemically, is the job of the *endocrine system*. The *lymphatic system* works with the veins in draining fluid from tissues and helps defend the body against infection. The *skin*, the body's largest organ, encloses and protects all the body's systems.

CHOOSE YOUR OWN COLORS

1. SKELETAL
2. MUSCULAR
3. CIRCULATORY
4. RESPIRATORY
5. NERVOUS
6. DIGESTIVE
7. URINARY
8. REPRODUCTIVE
9. ENDOCRINE
10. LYMPHATIC
11. SKIN

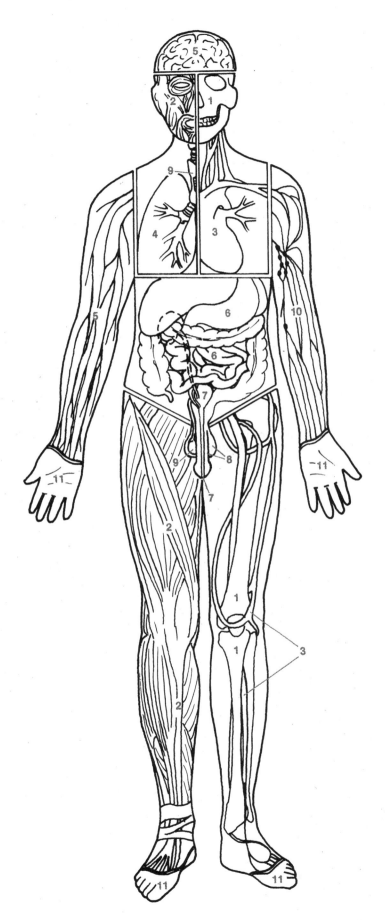

SKELETAL SYSTEM

The skeletal system in the adult consists of 206 *bones* and the strong elastic tissue that forms *ligaments, tendons,* and *cartilages,* which tie bones together and form the nose, larynx, trachea, bronchial tubes, and the outer ear. The skeleton provides a strong framework for the body, gives it its basic shape, and permits us to stand upright. The skeletal system also supports and restrains soft internal organs and shields fragile organs such as the brain and lungs. Certain bones, connected by flexible joints, form a combination of levers that allow coordinated movement. Bones also provide a firm anchor for skeletal muscles and produce red blood cells in their marrow cavities.

BONE CLASSIFICATION

Long Bones. These bones, such as those in the legs, arms, toes, and fingers, are strong shafts made of compact bone tissue. Their ends are large and consist of spongy tissue covered with compact tissue. They are slightly curved, enabling them to absorb shock.

Short bones. Shaped like irregular cubes, the short bones are spongy with a covering of compact tissue. The kneecap and the bones of the wrist and ankle belong to this category.

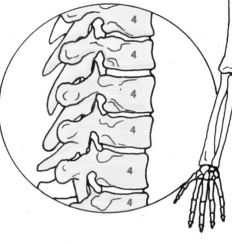

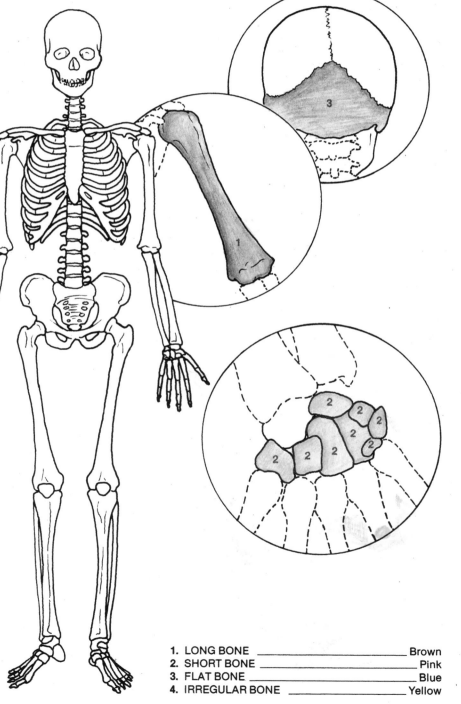

Flat bones. The skull, ribs, sternum, hips, and scapula are flat bones— bones with broad flat plates of spongy tissue sandwiched between two layers of compact tissue. Flat bones protect organs and are anchor points for muscles.

Irregular bones. As their name implies, these bones are irregularly shaped. The proportion of spongy to compact tissue varies from bone to bone. The vertebrae and facial bones belong to this group. Other irregular bones are put to special purposes, including helping to support and protect the body.

1. LONG BONE _____ Brown
2. SHORT BONE _____ Pink
3. FLAT BONE _____ Blue
4. IRREGULAR BONE _____ Yellow

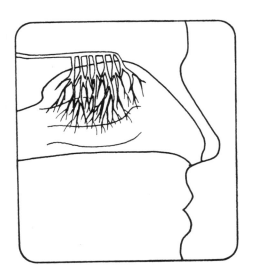

HUMAN ANATOMY COLORING BOOK

MARGARET MATT

TEXT BY JOE ZIEMIAN

Scientific advisor:
DONALD WERNSING, M.D.
Associate Director
Family Practice Residency Program
Summit, New Jersey

DOVER PUBLICATIONS, INC.
NEW YORK ·

Copyright

Copyright © 1982 by Margaret Matt and Joe Ziemian
All rights reserved.

Bibliographical Note

Human Anatomy Coloring Book is a new work, first published
by Dover Publications, Inc., in 1982.

International Standard Book Number
ISBN-13: 978-0-486-24138-8
ISBN-10: 0-486-24138-6

Manufactured in the United States by LSC Communications
24138641   2020
www.doverpublications.com

SKULL

The skull is the protective case for the brain and the organs of sight, taste, smell, hearing, and balance. It rests and pivots on the upper or superior end of the vertebral column. The skull has two main parts: the *cranium* or brain case and the *facial bones*. The base of the skull is much thicker and stronger than the sides and top and has many openings for nerves, blood vessels, and tubes to pass through. The facial bones enclose the front of the brain and form the openings for the eyes and the nasal and oral cavities. The *mandible* or jawbone is the only movable bone of the skull.

As the fetus develops, the cartilaginous membranes of the cranium *ossify* or turn into bone. At birth the ossification is not complete and membrane-filled spaces between the bones, the *fontanelles*, remain as soft spots. The largest, between the *parietal* and *frontal bones*, closes after about eighteen months.

1. FRONTAL BONE _____ Pink
2. PARIETAL BONE _____ Turquoise
3. SPHENOID BONE _____ Gray
4. ETHMOID BONE _____ Brown
5. LACRIMAL BONE _____ Green
6. NASAL TURBINATES _____ Red
7. VOMER_____ Light Blue
8. TEMPORAL BONE _____ Blue
9. NASAL BONE _____ Light Purple
10. ZYGOMATIC BONE _____ Orange
11. MAXILLA _____ Yellow-Green
12. MANDIBLE_____ Gray
13. TEETH _____ Yellow
14. OCCIPITAL BONE _____ Light Green
15. HYOID BONE _____ Light Brown

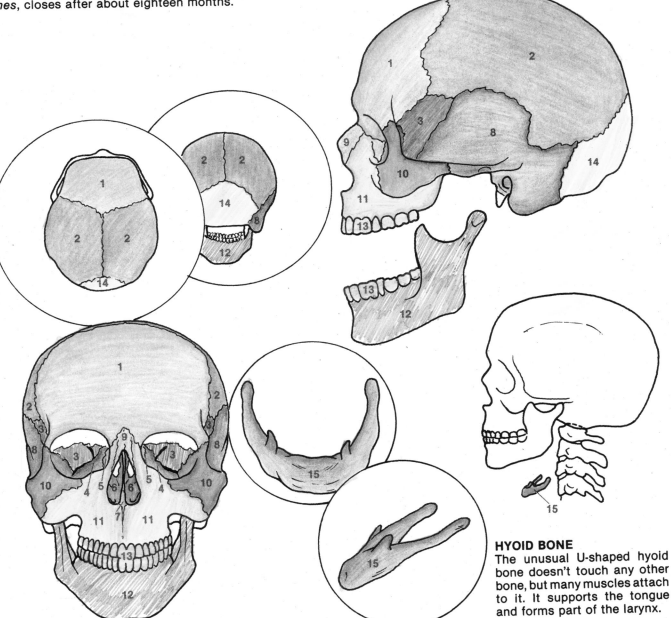

HYOID BONE
The unusual U-shaped hyoid bone doesn't touch any other bone, but many muscles attach to it. It supports the tongue and forms part of the larynx.

VERTEBRAL COLUMN — SPINE

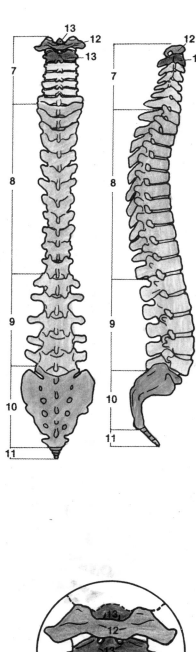

A stack of 33 irregular *vertebrae* or bones, all tied tightly together by ligaments, forms the strong, flexible column known as the spine. It can be divided into five regions. The *cervical* area forms the neck and supports the skull. The *thoracic* region, together with the ribs, forms the *thorax* or chest. The *lumbar* section makes up the lower back. The *sacrum* consists of five vertebrae fused into one triangular bone and forms part of the hip. The bottom four vertebrae, a remnant of the tail humans lost in evolutionary history, comprise the *coccyx*.

The lumbar vertebrae are the largest and thickest; the cervical are the smallest. All vertebrae have a thick body to bear weight and two wing-like *lamina* that join and form a ring, the *vertebral arch*. The ring opening is called the *vertebral foramen*. The openings are placed together to form an armored tube for the spinal cord, the *vertebral* or *spinal canal*. Each vertebra has seven *processes* or fingers that serve as anchors for muscles, contact points for vertebrae above and below, and overlapping shields to protect the spinal cord.

As a child grows, curves develop in the spine that give it strength and spring. These curves, together with the *disks* or cartilage pads between the vertebrae, protect the vertebral column by absorbing shock and concussion. The painful or numbing condition known as a *pinched nerve* occurs (usually in the lower back and occasionally in the neck) when a disk is crushed or ruptured in an accident or from lifting a heavy weight. The flattened or ruptured disk presses against or "pinches" nerves where they branch off from the spinal cord.

1. SKULL _____ Gray
2. STERNUM _____ Green
3. RIBS _____ Yellow
4. VERTEBRAL COLUMN _____ Light Purple
5. INTERVERTEBRAL DISKS _____ Flesh
6. HYOID BONE _____ Light Brown
7. CERVICAL VERTEBRAE _____ Light Blue
8. THORACIC VERTEBRAE _____ Purple
9. LUMBAR VERTEBRAE _____ Pink
10. SACRUM _____ Red
11. COCCYX _____ Orange
12. ATLAS _____ Blue
13. AXIS _____ Dark Blue
14. VERTEBRA BODY _____ Light Orange
15. TRANSVERSE PROCESS _____ Yellow-Green
16. SPINOUS PROCESS _____ Light Green
17. **a.** SUPERIOR and **b.** INFERIOR
    ARTICULAR PROCESS_____ Turquoise
18. COSTAL CARTILAGE _____ Brown

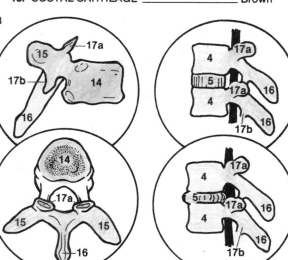

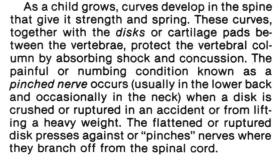

AXIAL SKELETON

The axial skeleton consists of the *skull*, the *vertebral column*, the *sternum*, and the *thorax* or rib cage and serves to hold the body erect. It also protects most of the body's vital organs such as the heart, lungs, and liver, which the thorax encloses. The thorax is a bony and somewhat flexible cage made up of twelve pairs of thin, curved ribs, the head of each of which joins or articulates with one or two vertebrae. The first seven pairs of ribs, the "true" ribs, are directly attached to the sternum by a strip of costal cartilage. The other five pairs, the "false" ribs, consist of the eighth, ninth, and tenth pairs, which are attached to each other and the seventh pair by cartilage, and the "floating" ribs (pairs eleven and twelve), which are not tied to the sternum at all but to the muscles of the abdominal wall. The elasticity of the cartilage and the flexible joints at the spine allow the ribs to flex in and out, reducing or enlarging the volume of the thorax.

The sternum develops in the child as three separate bones, but the three fuse into one in the adult around age 25. Pushing down on the lower third of the sternum compresses the heart and creates a pumping action that is the basis of cardiopulmonary resuscitation.

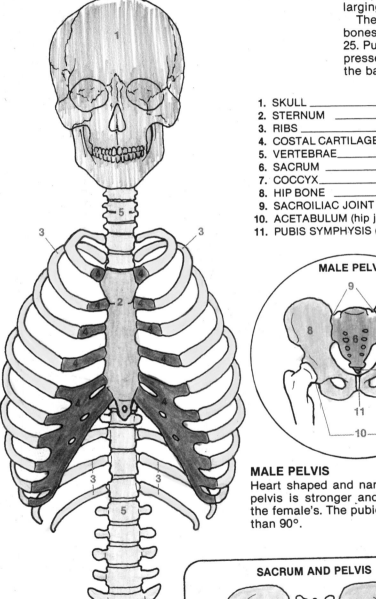

1. SKULL _____ Gray
2. STERNUM _____ Green
3. RIBS _____ Yellow
4. COSTAL CARTILAGE _____ Brown
5. VERTEBRAE_____ Light Purple
6. SACRUM _____ Red
7. COCCYX_____ Orange
8. HIP BONE _____ Pink
9. SACROILIAC JOINT
10. ACETABULUM (hip joint)
11. PUBIS SYMPHYSIS (interpubic joint)

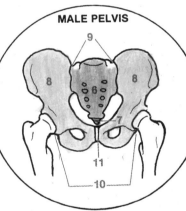

MALE PELVIS

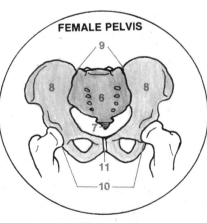

FEMALE PELVIS

MALE PELVIS

Heart shaped and narrow, the male pelvis is stronger and heavier than the female's. The pubic angle is less than 90°.

SACRUM AND PELVIS

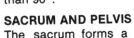

FEMALE PELVIS

Wider, smoother, and more bowl-like than the male's in order to accommodate the fetus during pregnancy and childbirth, the female pelvis is also smaller and structurally weaker than the male's. The pubic angle is greater than 90°.

SACRUM AND PELVIS

The sacrum forms a strong, interlocking keystone for the pelvis, which is subject to more stress than any other structure in the body.

APPENDICULAR SKELETON — UPPER EXTREMITIES

The combination of simple joints and levers that make up the arm and hand produce an astounding range of movements — baseball pitchers and jewelers both employ the complex arm and hand for very different purposes. The arm is supported by the *pectoral girdle*, which consists of the *scapula* or shoulder blade and the *clavicle* or collar bone. The scapula is held in place only by muscles; hence it is free floating and capable of considerable movement.

There is a ball at the top of the upper arm bone or *humerus* that rotates in a socket in the scapula. The forearm consists of the *ulna* and *radius*, which join the hand at the eight wrist bones or *carpals; intercarpal ligaments* tie the carpals together. The *metacarpals*, the five long bones that form the palm of the hand, join with the carpals. Beginning with the thumb, the metacarpals are numbered 1 to 5. The *knuckles* are the *heads* of the metacarpals. The finger bones or *phalanges* articulate with the metacarpals. Each finger has three bones, except for the thumb, which has two.

1. CLAVICLE _____ Green
2. **a.** SCAPULA, **b.** ACROMIAL PROCESS,
   and **c.** CORACOID PROCESS _____ Pink
3. HUMERUS _____ Purple
4. RADIUS_____Turquoise
5. ULNA _____ Gray
6. CARPALS _____ Yellow
7. METACARPALS _____ Brown
8. PHALANGES_____ Blue

FOREARM ROTATION

The forearm has an interesting mechanical arrangement that permits it to rotate and gives it power. The ulna is a stationary axle; the radius turns around it. Rotate your hand and you will see that the ulna doesn't move. To appreciate the power of forearm rotation, all you need do is try to prevent rotation by grasping your left wrist with your right hand. You will have to exert a great deal of force with your right hand to stop your left forearm from turning.

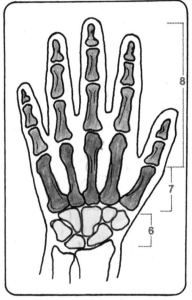

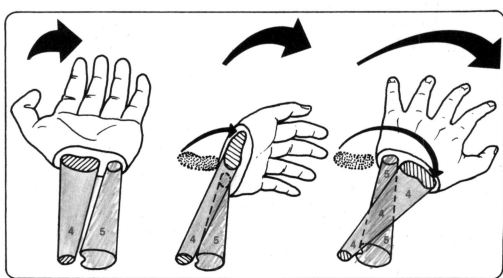

6

APPENDICULAR SKELETON — LOWER EXTREMITIES

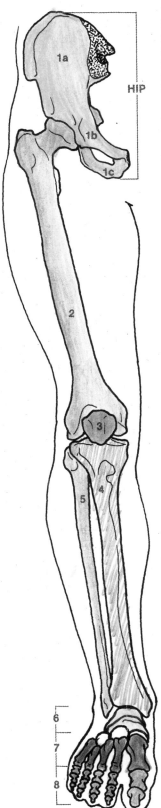

The bones of the lower and upper extremities are similar in many respects, but they serve different functions and, hence, have some structural differences. The leg bones must support the body's weight and are thus more solid than the arm bones, but the leg bones have a smaller range of movement.

At birth the hip has three bones—the *ilium, pubis,* and *ischium*—which later unite into a single bone, the *os coxa.* The left and right coxae join in the front at the *symphysis pubis* and with the *sacrum* in the back to form the bowl-shaped *pelvis.* The hip bone is connected to the thigh bone or *femur,* which has a ball-like head that rotates in the hip socket and a slight inward curve that aligns the body vertically with the knees and ankles. This alignment is important for the body to maintain its center of gravity. The lower femur and the *tibia* or shin form the knee. The small bone called the *patella* protects the knee; it is held in place by a tendon and surrounded by a *bursa,* a sac filled with fluid. Parallel to and outside of the tibia is the *fibula* or calf bone, whose lower end forms the outer ankle bone or *lateral malleolus.* The tibia and fibula articulate with the *talus,* the uppermost of the seven *tarsal bones.* The tarsals and the five *metatarsals* (numbered 1 to 5, beginning with the big toe) form two arches that act as a spring, distributing weight and helping to balance the body. The *phalanges* of the foot are similar to those of the hand in number and arrangement—two phalanges for the big toe, three for each of the other toes.

The condition known as fallen arches or "flat feet" results from the weakening of the ligaments and tendons that hold up the arches.

1. HIP: **a.** ILIUM, **b.** PUBIS, and
   c. ISCHIUM _____ Pink
2. FEMUR _____ Purple
3. PATELLA_____ Orange
4. TIBIA_____ Gray
5. FIBULA _____ Turquoise
6. TARSALS _____ Yellow
7. METATARSALS _____ Brown
8. PHALANGES _____ Blue

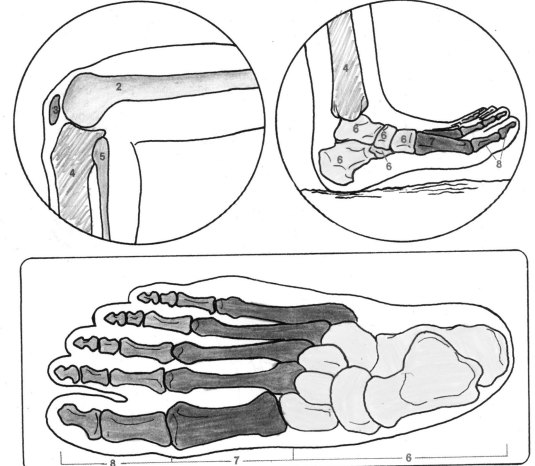

MUSCULAR SYSTEM

Muscles are tissues that contract, and the human body has almost seven hundred of them, which are divided into three kinds. *Skeletal muscles* are responsible for the voluntary movement of the bones. *Smooth muscles* are involuntary; they include blood vessels, intestines, and the lungs. There is only one *cardiac muscle* — the heart.

The term *muscular system* is used only for the skeletal muscles, which are the long, slender fibers arranged in parallel bundles that give our arms, legs, torso, neck, and face much of their shape. The large part of the muscle is called the *belly*. The ends of the skeletal muscles are attached by ligaments to two different bones, only one of which moves when the muscle contracts. The *origin* is where the muscle meets the bone that doesn't move. The origin is always closer to the torso than the *insertion*, where the muscle meets the bone that does move. Muscles that bend joints and pull limbs toward the body are called *flexors*. Muscles that straighten joints are *extensors*.

Movement usually involves the coordinated action of several muscles. The muscle that initiates the action is the *agonist* or *prime mover*. As the agonist contracts, another muscle, the *antagonist*, relaxes or yields to it. Other muscles, *synergists* or *fixators*, help the prime mover by dampening unwanted movement or holding a limb or joint steady during the action.

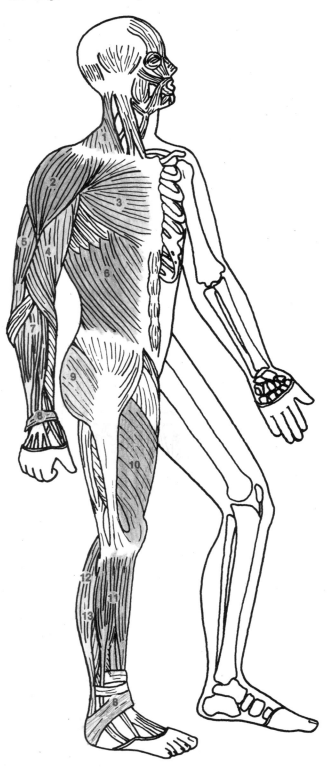

1. TRAPEZIUS_____ Pink
2. DELTOID _____ Orange
3. PECTORALIS _____ Light Blue
4. BICEPS _____ Green
5. TRICEPS _____ Turquoise
6. EXTERNAL OBLIQUE _____ Light Brown
7. EXTENSORS _____ Yellow
8. ANNULAR LIGAMENT_____ Gray
9. GLUTEUS MAXIMUS_____ Purple
10. QUADRICEPS _____ Red
11. PERONEUS_____ Blue
12. GASTROCNEMIUS _____ Pink
13. SOLEUS _____ Yellow-Green
14. TENDON _____ Flesh
15. ORIGIN OF MUSCLE
16. INSERTION OF MUSCLE

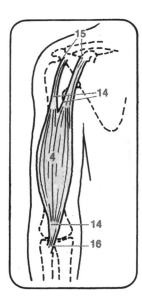

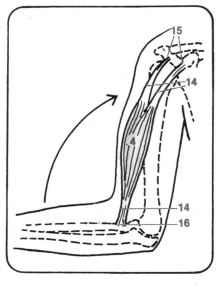

MUSCLES OF THE FACE, HEAD, AND NECK

The complex muscles of the head and neck are capable of rotation and powerful movement as well as the minute coordinated actions that express slight emotional changes in the face. Unlike most skeletal muscles, the face and head muscles are not attached to a moving limb. Instead they insert either into the flat bones of the skull or face or into head tissue such as the lip or skin of the chin. The *muscles of facial expression*, located near the skin (superficial muscles), register emotion and also help you chew and speak. The *muscles of mastication* move the lower jaw primarily for chewing, but they also are necessary for speaking. The *muscles of the tongue* help with chewing and swallowing and are extremely important for making the complex movements required for human speech. Some of the tongue muscles have both their origin and insertion at the hyoid bone; others insert into the tongue. The *muscles of the neck* arise primarily from the sternum and clavicle and as far down as the sixth vertebra. Neck muscles permit you to rotate and extend your head.

1. FRONTALIS _____ Orange
2. ORBICULARIS OCULI _____ Light Blue
3. TEMPORALIS _____ Purple
4. COMPRESSOR NARIS _____ Pink
5. LEVATOR LABII SUPERIORIS _____ Light Green
6. ZYGOMATICUS _____ Yellow
7. ORBICULARIS ORIS _____ Blue
8. BUCCINATOR _____ Light Brown
9. MASSETER _____ Red
10. MENTALIS_____ Brown
11. DEPRESSOR LABII INFERIORIS _____ Green
12. TRIANGULARIS_____ Turquoise
13. a. DIGASTRICUS, b. STERNOHYOIDEUS, and c. OMOHYOIDEUS _____ Yellow Green
14. a. STERNOCLEIDOMASTOIDEUS and b. TRAPEZIUS _____ Light Orange
15. SUPERIOR OBLIQUE_____ Gray
16. a. SUPERIOR, b. LATERAL, and c. MEDIAL RECTUS _____ Light Purple
17. TROCHLEA (pulley) _____ Dark Blue
18. INFERIOR OBLIQUE_____ Flesh

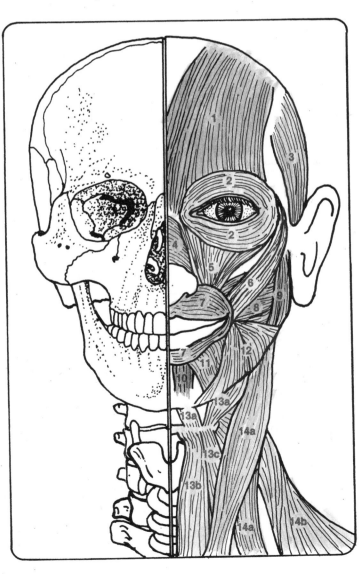

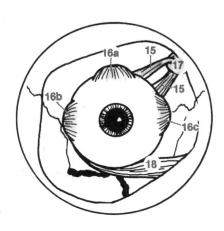

MUSCLES OF THE TORSO

The muscles that encase the torso hold the body erect, allow it to flex, assist in breathing, and restrain the intestines. The *back muscles* that control the torso's forward and backward movement originate primarily along the upper crest of the hip, the sacrum, and the lumbar vertebrae and rise vertically and diagonally into the ribs and vertebrae at various levels. Although the lumbar vertebrae can flex freely in all directions, the movement of the thoracic vertebrae and their correspondent ribs is very limited, which insures that the lungs are not compressed and have room to expand during breathing. A web of large muscles that restrains the abdominal contents, bends the spine, and assists in breathing passes vertically, horizontally, and diagonally across the abdominal cavity. For example, *obliquus externus* (external oblique) compresses the abdominal cavity during forced exhalation. If the external oblique muscle on only one side of the body contracts, the spine bends sideward. The muscles used in breathing enlarge and reduce the size of the thoracic cavity, which decreases and increases in turn the air pressure in the lungs. The *diaphragm,* which forms the floor of the thoracic cavity, contracts downward to increase the length and therefore the volume of the thorax. Two sets of *intercostal muscles* placed side by side fill the spaces between the ribs. The external or outer intercostals, which raise the ribs and enlarge the cavity during inspiration, angle downward and forward away from the spine. The internal or inner set are cast down and back toward the spine and serve to draw the ribs down to reduce the volume of the thoracic cavity for exhalation.

1. DELTOID _____ Orange
2. PECTORALIS MAJOR _____ Light Blue
3. LATISSIMUS DORSI _____ Red
4. SERRATUS ANTERIOR _____ Gray
5. EXTERNAL OBLIQUE_____ Light Brown
6. RECTUS ABDOMINIS _____ Turquoise
7. INTERNAL OBLIQUE _____ Brown
8. EXTERNAL INTERCOSTAL_____ Yellow
9. INTERNAL INTERCOSTAL _____ Green
10. TRAPEZIUS_____ Pink
11. TERES MAJOR _____ Yellow-Green

MUSCLES OF THE UPPER LIMBS

A complex group of strong opposing muscles move the arm. These muscles, which arise from the scapula, clavicle, sternum, ribs, lower vertebrae, and hips, form the only attachment between the arm and shoulder and the axial skeleton and hold the humerus in its socket. By contracting and relaxing in combination, the shoulder muscles are able to rotate, extend, and flex the arm at the shoulder. The *brachialis, biceps brachi* (which has two *heads* or origins), and *triceps brachi* (three heads), all muscles of the upper arm, flex the elbow joint and move the forearm. The two rotating actions, *supination* (as when you turn a key) and *pronation* (as in turning the palm down) are generated by muscles that arise in the humerus and wrap around the radius and ulna like a window shade around its roller. The forearm and lower humerus are the origin for the primary muscles of the wrist, hand, and fingers. The fingers are connected by long tendons, which you can see in the back of your hand, that run from the forearm muscles. Muscles on the underside of the forearm bend the fingers; muscles on the upper side extend them. The hand has small muscles that spread the fingers and perform the complex and very important apposable thumb action, which man alone among primates is capable of.

1. DELTOID _____ Orange
2. PECTORALIS MAJOR _____ Light Blue
3. TRICEPS _____ Pink
4. BICEPS _____ Green
5. BRACHIALIS _____ Light Brown
6. PRONATOR TERES _____ Gray
7. BRACHIORADIALIS _____ Light Purple
8. FLEXOR CARPI RADIALIS _____ Yellow
9. PALMARIS LONGUS _____ Turquoise
10. FLEXOR CARPI ULNARIS _____ Red
11. FLEXOR DIGITORUM SUPERFICIALIS _____ Light Green
12. EXTENSOR DIGITORUM and INDICIS _____ Light Orange

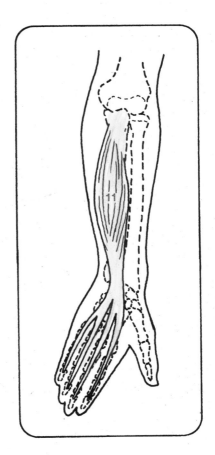

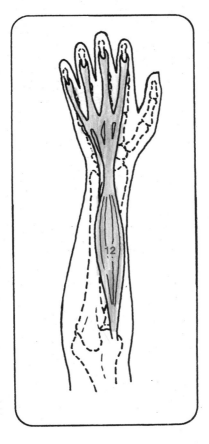

MUSCLES OF THE LOWER LIMBS

The leg is far more powerful than the arm, but its motion is more limited. The upper leg is moved and rotated by large muscles that arise from the front, side, and back of the hip and sacrum. The *gluteus* or *buttock muscles* principally extend, rotate, and elevate the femur. The thigh muscles move the knee. Four muscles, *quadriceps femoris*, which form the front of the thigh, join in a strong tendon just above the knee. The tendon inserts at the top of the tibia; the patella is embedded in it. The *sartorius*, the longest muscle of the body, arises in the ilium, crosses the quadriceps, and inserts into the inner side of the tibia. Running down the back of the thigh is a group of muscles, the *hamstrings*, that extend and flex both the knee and hip. You can feel their large tendons under your knee when you sit down.

The calf muscles rotate and flex the foot and extend the toes. The lower leg muscles have long tendons that pass over the ankle and insert into the metatarsals and phalanges. Although the foot is relatively fragile, it can withstand strong forces through the progressive distribution of the body's weight from the heel to the shock-absorbing arch and ball. When you walk, the *dorsiflexors* rotate your foot upward, concentrating the majority of your weight on the heel bone as it strikes the ground. Two calf muscles, *gastrocnemius* and *soleus*, lift the heel through the *achilles tendon (tendocalcaneus)* and roll the foot forward, thus distributing your weight onto the metatarsal heads and toes. The foot has a number of small muscles that help support the toes and balance the body.

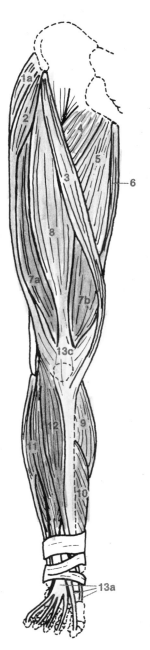

1. **a.** GLUTEUS MEDIUS and **b.** MAXIMUS _____ Purple
2. TENSOR FASCIAE LATAE _____ Flesh
3. SARTORIUS _____ Yellow-Green
4. PECTINEUS _____ Gray
5. ADDUCTOR LONGUS _____ Light Green
6. GRACILIS _____ Pink
7. **a.** VASTUS LATERALIS and **b.** MEDIALIS _____ Red
8. RECTUS FEMORIS _____ Light Blue
9. GASTROCNEMIUS _____ Green
10. SOLEUS _____ Orange
11. PERONEUS _____ Blue
12. TIBIALIS ANTERIOR _____ Brown
13. **a.** TENDONS, **b.** TENDOCALCANEUS, and
    c. TENDON OF QUADRICEPS FEMORIS _____ Yellow
14. **a.** SEMIMEMBRANOSUS and **b.** SEMITENDINOSUS __ Turquoise
15. BICEPS FEMORIS _____ Light Purple
16. EXTENSOR DIGITORUM BREVIS _____ Light Brown
17. ABDUCTOR DIGITI MINIMI _____ Dark Blue
18. ABDUCTOR HALLUCIS _____ Dark Green
19. FLEXOR DIGITORUM BREVIS _____ Light Orange

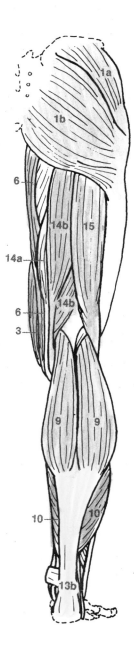

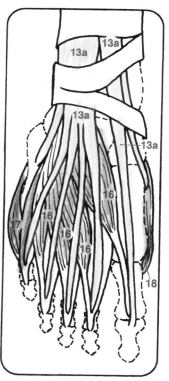

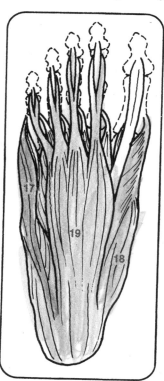

CIRCULATORY SYSTEM

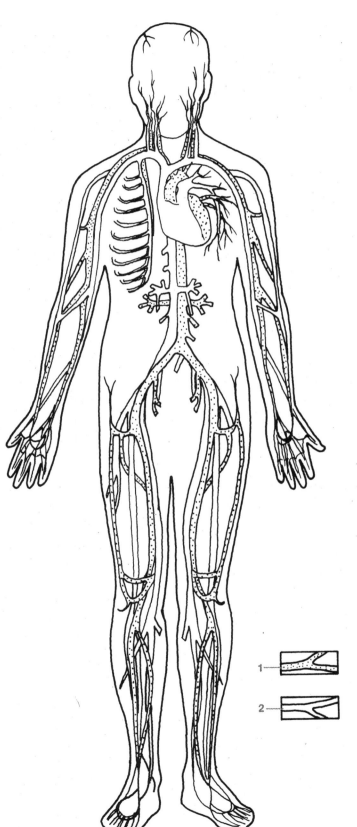

The circulatory system supplies oxygen and nutrients to every cell of the body and removes wastes and carbon dioxide. The system consists of *blood*, which carries the nutrients and wastes; the *heart*, which pumps the blood; and a closed system of tubes (*arteries* and *veins*) that carries the blood to and from the body tissues. The arteries transport blood enriched with oxygen and nutrients; the veins carry depleted blood. The artery that leaves the heart is very large, but it divides again and again into smaller and smaller branches. The tiniest of these branches are called *capillaries*, which are only seven to nine microns wide—so small that blood cells must pass through in single file. The capillaries are the site of the exchange of nutrients and wastes between the blood and the tissue cells. Interlacing capillary beds are found throughout the body except in the cartilages, cuticles, nails, hair, and the cornea of the eye. The depleted blood moves back toward the heart through the *venous system.* First the tiniest blood vessels unite in the capillary beds to form *venules,* then the venules combine again and again until they form the largest veins. The heart pumps the depleted blood to the lungs, where carbon dioxide is exchanged for oxygen, and to the liver and kidneys, which remove wastes.

Large arteries provide direct "express" service to major areas of the body such as the brain, lungs, arms, and abdomen; these arteries don't begin to subdivide until they reach the appropriate area. Press your finger against one of these arteries and you will feel the pump stroke or beat of the heart. This is the pulse.

1. ARTERIAL CIRCULATION _____ Red
2. VENOUS CIRCULATION _____ Blue

HEART

The heart is the key organ of the circulatory system. It is a bit larger than a man's fist — about twelve centimeters long, nine centimeters wide, and six centimeters thick. This hollow muscle is located to the left of the sternum between the second and fifth ribs and is enclosed in the *pericardium*, a membranous sac with a fibrous layer outside and a serous one inside, which protects the heart and anchors it in place. Between the two layers is a watery lubricant that minimizes friction when the heart beats. The heart is surrounded by the lungs, each of which has a notch, the *cardiac impression*, the heart fits into.

Each half of the heart has two chambers, the *atrium* (upper) and the *ventricle* (lower). Blood returning to the heart enters the right atrium from three veins: the *superior vena cava*, which runs from the upper torso and limbs; the *inferior vena cava*, which carries blood from the lower torso and limbs; and the *coronary sinus*, which circulates venous blood from the walls of the heart. The blood is pumped through the *tricuspid valve* (which has three cusps or flaps) into the right ventricle. From there it goes to the *pulmonary artery,* the only artery that carries unoxygenated blood, which carries it to the lungs. In the lungs the blood exchanges carbon dioxide for oxygen. The enriched blood then goes to the left atrium, where it passes through the strong *bicuspid* or *mitral valve* into the left ventricle. Finally the blood leaves the heart through the *aortic semilunar valves* and flows into the *aorta* and through the body.

The heart beats involuntarily, that is, the brain doesn't have to command it to pump blood. Instead various sensors monitor the body's activities and the consequent demands for more or less blood. For example, there are pressure receptors in the aorta that respond to changes in arterial pressure. The *aortic reflex* slows the heart when the pressure gets too high; the *carotid sinus reflex* increases the heart rate when the pressure becomes too low in the arteries that serve the brain. Chemoreceptors increase the heartbeat if they detect a lack of oxygen or an increase of carbon dioxide.

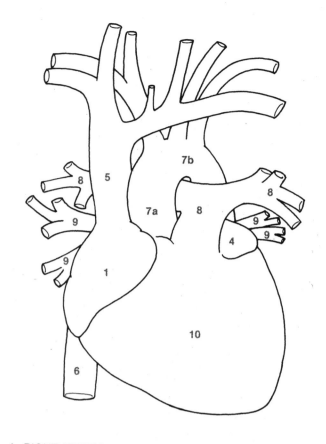

1. RIGHT ATRIUM _____ Light Purple
2. RIGHT VENTRICLE _____ Blue
3. LEFT VENTRICLE _____ Red
4. LEFT ATRIUM _____ Orange
5. SUPERIOR VENA CAVA _____ Light Blue
6. INFERIOR VENA CAVA _____ Dark Blue
7. **a.** ASCENDING AORTA and **b.** AORTIC ARCH_____ Pink
8. PULMONARY ARTERY _____ Green
9. PULMONARY VEIN _____ Yellow
10. HEART _____ Purple

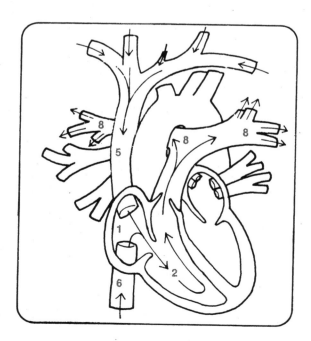

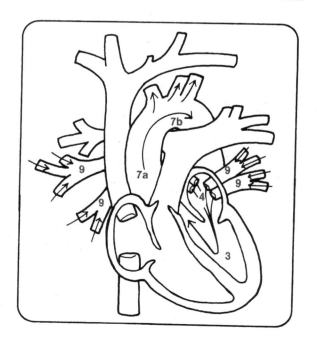

PULMONARY CIRCULATION

Pulmonary circulation is the movement of blood from the heart to the air sacs in the lungs and back to the heart. This circulation is necessary for the blood to exchange carbon dioxide for oxygen. The right ventricle of the heart pumps unoxygenated blood into the *pulmonary artery*, which divides in two to form the *right* and *left pulmonary arteries*. The right artery, which supplies the larger lung, the right (which has three lobes), is wider and longer than the left artery (which services the two-lobed left lung). The arteries enter the lungs at the *hilus*, a vertical slit. Then the arteries divide and pass through each lobe. Subdivision follows subdivision until massive capillary beds are formed that surround the *alveoli* or small air sacs. At this point there is only a thin membrane separating each capillary from the air sac. As a blood cell passes down a capillary, carbon dioxide diffuses through the membrane into the alveoli; then oxygen passes from the alveoli to the blood cell. The oxygenated blood moves out of the capillary bed into the increasingly larger pulmonary veins. These veins unite to form four main trunks (two for each lung) and empty into the left atrium of the heart. The heart then pumps the oxygenated blood on its path through the body.

1. RIGHT ATRIUM _____ Light Purple
2. RIGHT VENTRICLE _____ Blue
3. LEFT VENTRICLE _____ Red
4. LEFT ATRIUM _____ Orange
5. SUPERIOR VENA CAVA _____ Light Blue
6. INFERIOR VENA CAVA (opening) _____ Dark Blue
7. **a.** ASCENDING AORTA and **b.** ARCH of AORTA _____ Pink
8. PULMONARY ARTERY _____ Green
9. PULMONARY VEIN _____ Yellow
10. LUNG _____ Gray
11. HEART _____ Purple
12. DIAPHRAGM _____ Brown
13. RIBS _____ Light Brown

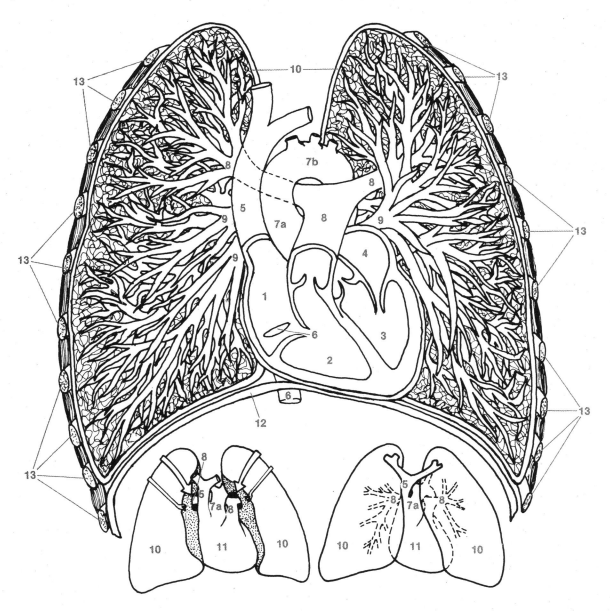

ARTERIES

The arteries carry blood from the heart to the capillaries, dividing and then subdividing on the way until they become smaller and smaller and more and more numerous, and end in the *capillary beds* located in the body tissues. The *aorta* is the first and largest of the arteries. Adults have about seven thousand square miles of capillaries. If laid end to end, all the elements of the *vascular* (blood vessel) *system*—arteries, capillaries, and veins—would extend about seventy thousand miles. The arteries have three layers—muscle tissue, elastic fibers, and connective tissue—and expand and contract in coordination with the flow of blood passing through them. Each heartbeat pushes blood into the arteries, which expand to hold the blood and then contract behind it as the heart pumps the blood to the next section of the vascular system. The arteries' structure prevents them from collapsing when broken, but the arteries will constrict to reduce the size of the opening and thereby diminish the loss of blood. Many parts of the body are served by more than one artery, a system called *collateral circulation*. Thus if a blood vessel serving such an area is damaged or restricted, the flow of blood will not stop completely.

1. HEART _____ Purple
2. **a.** ARCH OF AORTA, **b.** THORACIC AORTA, and
   c. ABDOMINAL AORTA _____ Pink
3. COMMON ILIAC _____ Orange
4. **a.** FEMORAL and **b.** TIBIAL _____ Light Brown
5. SUBCLAVIAN _____ Yellow
6. AXILLARY AND BRACHIAL _____ Green
7. **a.** ULNAR and **b.** RADIAL _____ Light Green
8. CAROTID _____ Light Orange
9. VERTEBRAL _____ Brown
10. ARTERIAL BLOOD _____ Red
11. CONNECTIVE TISSUE _____ Light Purple
12. SMOOTH MUSCLE TISSUE_____ Turquoise
13. SEROUS MEMBRANE _____ Flesh

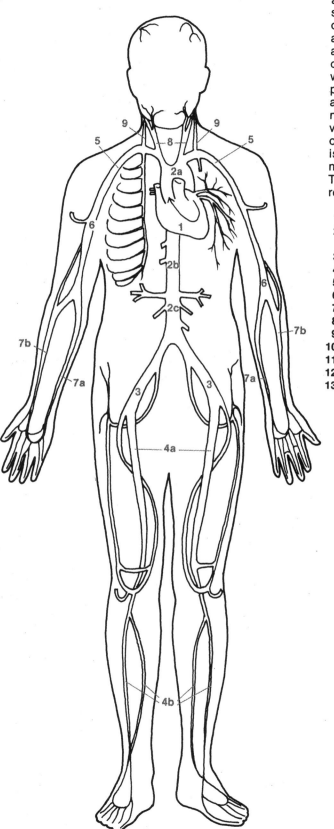

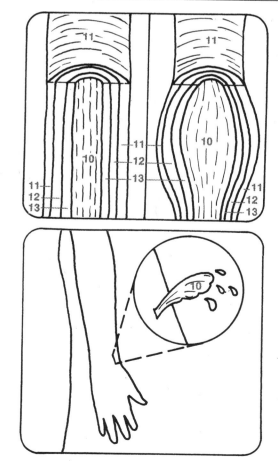

The veins bring depleted, unoxygenated blood from the capillary beds back to the heart. Unoxygenated blood is deep red. Blood turns bright red when it is exposed to oxygen in the air sacs of the lungs. The veins' structure is similar to that of the three-layered arteries, but the blood pressure in the veins is much less than in the arteries; thus the venous walls are thinner and less rigid and will collapse if they are not filled with blood. Although the venous walls are expandable, they are not as elastic as arterial walls. The pressure in the veins that pushes the blood back to the heart can barely overcome the pull of gravity. To prevent blood from flowing backward or pooling, veins most affected by gravity and the weight of the blood (veins of the lower leg, for example) have valves with double flaps that hold the blood until the pressure builds up sufficiently to move the blood toward the heart. When the pressure drops, the weight of the blood forces the valves shut. The condition known as *varicose veins* occurs, usually with age, when the valves weaken from constant stretching, which destroys the veins' elasticity, thereby permitting blood to leak downward and expand the walls of the veins.

Unlike arterial blood, venous blood does not spurt from a cut; rather it flows steadily. To stop the flow, apply pressure on the vein on the side of the cut furthest from the heart. The spurting blood from a cut artery must be stopped the opposite way, by applying pressure between the heart and the cut.

VEINS

1. HEART _____ Purple
2. SUPERIOR VENA CAVA _____ Light Blue
3. INFERIOR VENA CAVA _____ Dark Blue
4. COMMON ILIAC _____ Orange
5. **a.** GREATER and **b.** LESSER SAPHENOUS _____ Brown
6. **a.** FEMORAL and **b.** DEEP FEMORAL _____ Light Brown
7. SUBCLAVIAN _____ Yellow
8. CEPHALIC _____ Light Green
9. AXILLARY and BRACHIAL _____ Green
10. INTERNAL and EXTERNAL JUGULAR _____ Light Orange
11. LUNG _____ Gray
12. UNOXYGENATED BLOOD _____ Deep Red
13. OXYGENATED BLOOD _____ Bright Red
14. VALVES _____ Yellow-Green
15. SEROUS MEMBRANE _____ Flesh
16. SMOOTH MUSCLE TISSUE _____ Turquoise
17. CONNECTIVE TISSUE _____ Light Purple
18. RED CORPUSCLE IN CAPILLARY _____ Pink

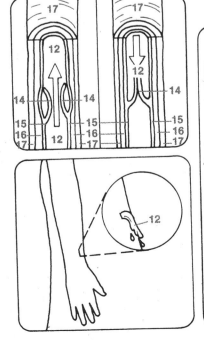

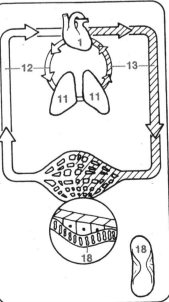

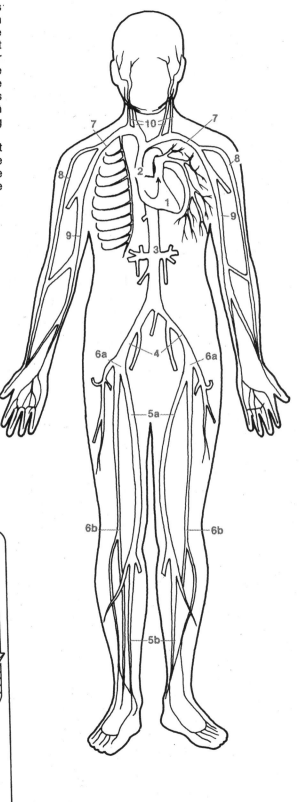

RESPIRATORY SYSTEM

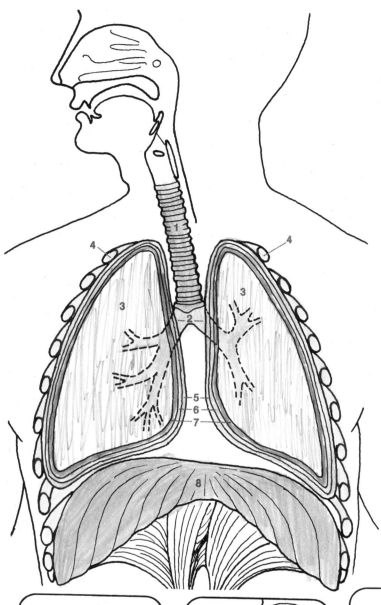

Every cell in the body converts oxygen to energy and generates carbon dioxide as a waste product. *Ventilation* is the term for the process of breathing oxygen in and carbon dioxide out. The respiratory system consists of the *nose*; *nasal cavities*, which filter and condition incoming air; the *pharynx* or throat; the *larynx* or voice box; the *trachea* or windpipe; the *lungs*; the *bronchi* or branching air tubes in the lungs; and the *air sacs*, the actual site of the oxygen–carbon dioxide exchange. The lungs are protected by the strong rib cage and, underneath, the diaphragm; they are surrounded by two serous membranes, the *visceral pleurae*. The lungs in turn surround the *mediastinum*, an interpleural space, that contains the heart in its pericardial sac and parts of the trachea, bronchi, esophagus, blood vessels, and nerves. The thoracic cavity is lined by a membrane called the *parietal pleura;* between it and the visceral pleurae is a potential space, the *intrapleural space*, containing only a thin fluid that acts as a lubricant for the pleurae, which thus slide frictionlessly as the lungs move. The lungs open their inner air chambers to the outside atmosphere; thus when the lungs are at rest the air pressure is the same inside and outside the body. As inhalation begins, the ribs, thoracic muscles, and diaphragm increase the size of the thoracic cavity, thereby lowering the air pressure in the lungs. The higher pressure outside the body then forces more air into the lungs to equalize the air pressure inside and outside. During exhalation the muscles and ribs compress the lungs, raising the air pressure inside until it exceeds the pressure outside — and the air rushes out.

1. TRACHEA _____ Blue
2. BRONCHUS _____ Light Blue
3. LUNG_____ Gray
4. RIB CAGE _____ Yellow
5. PARIETAL PLEURA _____ Pink
6. INTRAPLEURAL SPACE ____ Yellow-Green
7. VISCERAL PLEURA _____ Orange
8. DIAPHRAGM_____ Red
9. a. RIGHT SUPERIOR and
   b. LEFT SUPERIOR LOBES __ Light Orange
10. RIGHT MIDDLE LOBE _____ Light Green
11. a. RIGHT INFERIOR and
    b. LEFT INFERIOR LOBES ___ Light Brown
12. HEART _____ Purple
13. LUBRICATING FLUID_____ Green

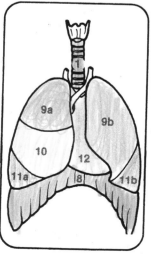

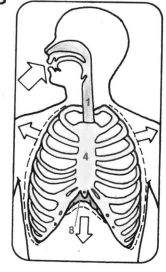

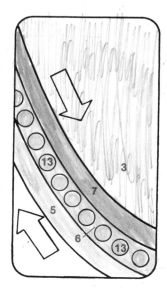

NOSE, NASAL CAVITIES, AND PHARYNX

The air you inhale passes through a series of chambers and passageways that cleans and conditions it before it reaches your lungs. First the air enters the body through the two *nostrils*, which open into two large chambers, the nasal cavities. The cavities are separated from each other by the *septum*, a cartilage divider. Sticky mucous membranes line the walls of the cavities. The *cilia*, millions of hair-like projections that cover the membranes, sweep trapped particles toward the throat to be swallowed. Each cavity also has a set of three curved bones, the *turbinates* or *conchae,* that serve as airfoils, causing air passing through them to swirl and change direction. Larger airborne particles that can't follow the airstream get thrown on the sticky membrane. The nasal cavities can trap almost all particles over ten microns wide. The nasal passages also warm incoming air almost to body temperature and humidify it by adding moisture from the mucous membranes and drainage from various sinuses. As much as 350 to 400 milliliters of water per day is used to humidify dry air. The nasal cavities are also protected by the *sneezing mechanism*, which is triggered when irritants stimulate a reaction from the brain to command the deep inhalation and explosive exhalation that carries away the foreign matter.

Both food and air pass into the body through the pharynx or throat, which is located just behind the nasal cavities and mouth and is made of skeletal tissue and lined with mucous membranes. It has three parts: the *nasopharynx* is on the top near the nasal cavities; the *oropharynx* is in the middle behind the oral cavity; and the *laryngopharynx* is located above the voice box and esophagus. The *eustachian* or *auditory tubes* open into the pharynx. They drain secretions from the middle ear and connect the middle ear, nasopharynx, and the atmosphere in order to equalize pressure on both sides of the eardrum.

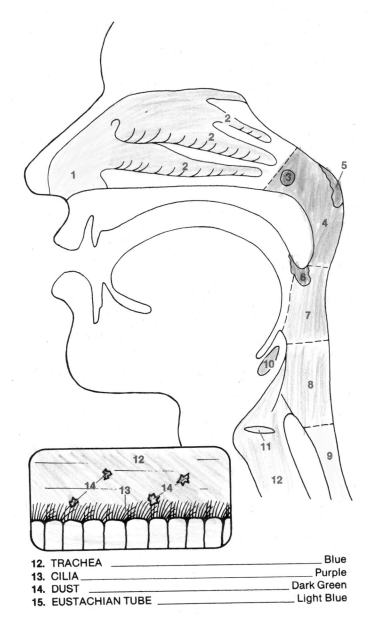

1. NASAL CAVITY _____ Light Green
2. TURBINATE BONES _____ Green
3. OPENING OF EUSTACHIAN TUBE _____ Dark Blue
4. NASOPHARYNX _____ Orange
5. ADENOIDS _____ Brown
6. TONSILS _____ Gray
7. OROPHARYNX _____ Light Orange
8. LARYNGOPHARYNX _____ Light Brown
9. ESOPHAGUS _____ Yellow-Green
10. EPIGLOTTIS_____ Red
11. LARYNX _____ Yellow

12. TRACHEA _____ Blue
13. CILIA _____ Purple
14. DUST _____ Dark Green
15. EUSTACHIAN TUBE _____ Light Blue

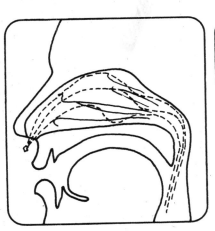

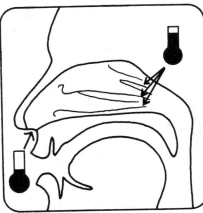

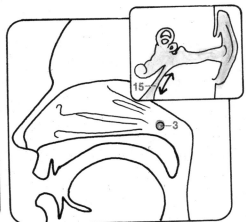

MECHANICS OF BREATHING AND SWALLOWING

The *laryngopharynx* is shared by air, solid foods, and liquids passing into the body. The *larynx* and the *esophagus* open into it. A type of mucous membrane, *stratified squamous epithelium*, coats its walls. Sensors in the epithelium act as traffic controllers that allow only air to enter the *trachea*. When solids or liquids stimulate these touch receptors, the *epiglottis*, a small flap over the trachea opening, snaps shut. At the same time swallowing begins to carry the material into the stomach.

1. SUPERIOR CONCHA _____ Light Green
2. MIDDLE CONCHA _____ Green
3. INFERIOR CONCHA _____ Dark Green
4. PHARYNX _____ Orange
5. **a.** EPIGLOTTIS and **b.** CORK _____ Red
6. LARYNX _____ Yellow
7. TRACHEA _____ Blue
8. ESOPHAGUS _____ Yellow-Green
9. HYOID BONE _____ Light Brown
10. **a.** AIR IN LUNGS and **b.** BALLOON Light Blue

COUGH SEQUENCE

Even minute particles can block air passages in the lungs. The walls of the trachea are lined with cilia and mucous membranes that trap dust not caught in the upper cavities. Large particles stimulate touch sensors that trigger the brain to send out a cough reaction signal. The lungs fill with air, the epiglottis closes, and the ribs and diaphragm are pushed sharply against the lungs, thereby increasing the lungs' pressure — it's like squeezing a balloon. The epiglottis flies open and air rushes out at almost the speed of sound, dislodging the particle and carrying it out.

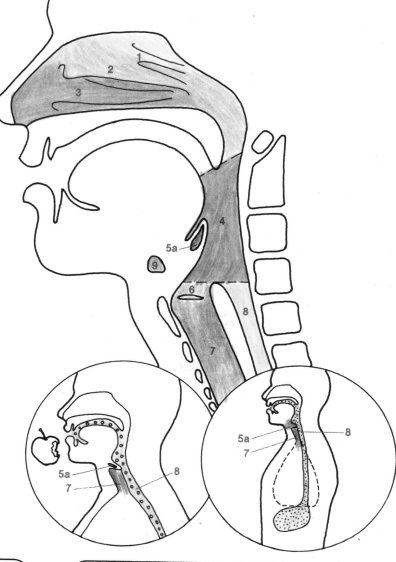

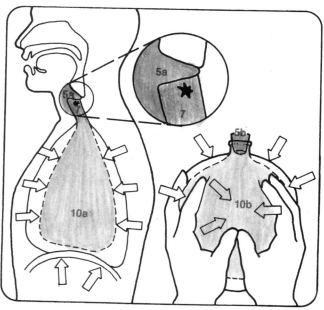

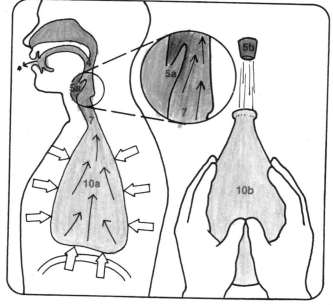

LARYNX AND TRACHEA

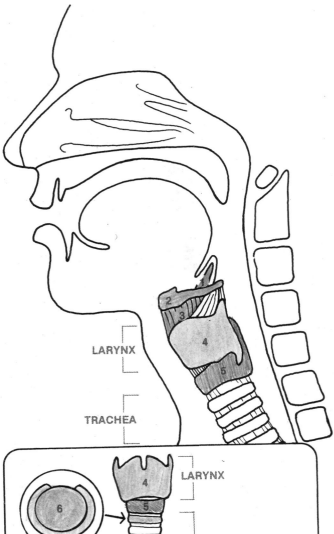

The larynx or voice box connects the pharynx with the top of the trachea. The larynx is made of nine cartilages, all tied together by ligaments and moved by muscles, and two pairs of mucous membrane folds. The lower or inferior folds are the *true vocal folds* or *cords;* the upper or superior folds are the *false vocal folds* or *cords*. Only the true cords create sound; they vibrate like violin strings in the airstream. The *glottis*, which consists of the true cords and the opening between them, permits a variable amount of air to pass through the voice box. The false folds prevent solids from entering the larynx, and they also come into play when you hold your breath.

The trachea or windpipe extends from below the larynx and toward the lungs. It is reinforced by a column of C-shaped cartilages and coated with mucous membranes and cilia that trap and sweep minute dust particles upward to the pharynx. The trachea divides into two *bronchi*, one for each lung, which are structurally similar to the trachea. The right bronchus is the wider, shorter, and more vertical of the two. The bronchi divide and subdivide continuously, becoming smaller and more numerous. As they become smaller they change from cartilage rings to cartilage plates to smooth muscles. These are the muscles that spasm during an asthma attack, closing off air passageways and making breathing difficult.

There is little change in velocity and pressure of the air as it moves through the bronchial system. This is because the number of tubes increases even as the size of each individual tube decreases. It takes just a little bit of muscular force to create a vacuum in the tubes for inhalation or high pressure for exhalation.

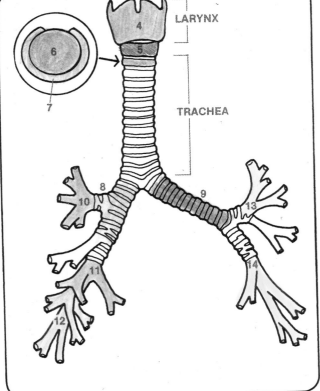

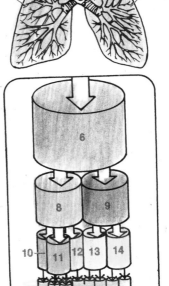

1. **EPIGLOTTIS** _____ Red
2. **HYOID BONE** ____ Light Brown
3. **LIGAMENTS** _____ Orange
4. **THYROID CARTILAGE** _____ Green
5. **CRICOID CARTILAGE** __ Gray
6. **TRACHEA** _____ Blue
7. **TRACHEAL CARTILAGE**_____Light Purple
8. **RIGHT PRIMARY BRONCHUS** _____ Light Blue
9. **LEFT PRIMARY BRONCHUS**_____Dark Blue
10. **RIGHT UPPER LOBAR BRONCHUS** _____ Dark Green
11. **RIGHT MIDDLE LOBAR BRONCHUS** ____ Pink
12. **RIGHT LOWER LOBAR BRONCHUS** ____ Yellow-Green
13. **LEFT UPPER LOBAR BRONCHUS** ____ Light Green
14. **LEFT LOWER LOBAR BRONCHUS** _____ Yellow
15. **BRONCHIOLES**____ Turquoise

ALVEOLI

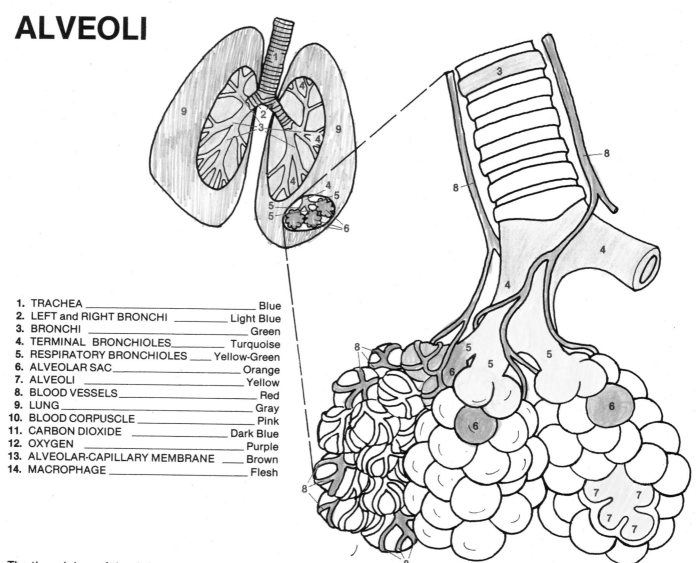

1. TRACHEA _____ Blue
2. LEFT and RIGHT BRONCHI _____ Light Blue
3. BRONCHI _____ Green
4. TERMINAL BRONCHIOLES_____ Turquoise
5. RESPIRATORY BRONCHIOLES ____ Yellow-Green
6. ALVEOLAR SAC_____ Orange
7. ALVEOLI _____ Yellow
8. BLOOD VESSELS_____ Red
9. LUNG _____ Gray
10. BLOOD CORPUSCLE _____ Pink
11. CARBON DIOXIDE _____ Dark Blue
12. OXYGEN _____ Purple
13. ALVEOLAR-CAPILLARY MEMBRANE ____ Brown
14. MACROPHAGE _____ Flesh

The three lobes of the right lung and the two of the left are all served by *secondary bronchi*. Like arteries, the bronchi divide again and again, becoming more numerous, narrower, and shorter with each subdivision. The final and smallest of the bronchial air tubes are the *terminal bronchioles*. Even smaller than the terminal bronchioles are another kind of tube, the *respiratory bronchioles*, which are made of a smooth layer of ciliated epithelium, and which turn into the *atrium*, an elongated sac-like opening. (*Epithelium* is the term for the cellular tissue covering an internal or external surface.) The alveoli or air cells, small projections that number about three hundred million in an adult, line the respiratory bronchioles. The enormous area of the alveoli (if spread out, the alveoli would almost cover a football field) is protected from airborne irritants and organisms and kept sterile by the *macrophages*, specialized cells in the alveolar wall that surround and digest unwanted particles smaller than one micron. In the alveoli only the thin *alveolar-capillary membrane* separates the air from the blood-carrying capillaries. When you breathe in, the air in the alveoli is 21 percent oxygen, .04 percent carbon dioxide, and 79 percent nitrogen; but the blood cells coming back to the lung have already given up some of their oxygen to tissues throughout the body and thus carry less oxygen and more carbon dioxide—a chemical imbalance on the two sides of the alveolar-capillary membrane. Thus the two gases try to equalize their pressures: oxygen passes from the alveoli through the membrane and to the blood, and carbon dioxide diffuses through the membrane to the air sac. The large area of the alveoli makes for a very efficient exchange: oxygen is replenished in about one-fifth of a second.

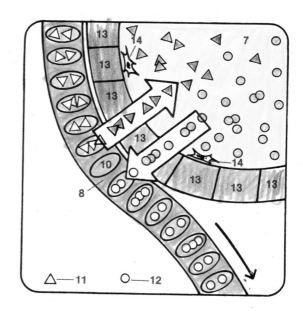

BREATHING MECHANISM

No conscious muscle force is required for *inspiration* (breathing in) or *expiration* (breathing out). Before the breathing cycle begins your lungs are at rest and the air pressure in them equals the pressure outside. Then the diaphragm contracts, which increases the vertical measure of the thorax. The central tendon pulls downward and flattens the diaphragm. The external intercostal muscles pull the ribs upward and outward; this increases the diameter of the rib cage. During deep inhalation the neck and back muscles help elevate the ribs. As a result of all this, the lungs are pulled by the parietal and visceral pleurae and expand — it's like having your hands glued to a balloon and pulling it wider. The air molecules inside the lungs now must fill a larger volume, which reduces the air pressure. The relatively higher pressure of air outside the body forces air into the body's airways and the lungs in order to equalize the pressure inside and outside.

During expiration the diaphragm and external intercostals relax and the elastic lungs attempt to spring back to the smaller size they were before inspiration. As the air space in the lungs gets smaller, the pressure in the lungs increases until it exceeds the air pressure outside the body. The cycle ends when the air rushes out of the lungs to equalize the pressure within and without, carrying with it the waste gas carbon dioxide.

1. **RIBS** _____ Yellow
2. **COSTAL CARTILAGE** _____ Yellow-Green
3. **STERNUM** _____ Green
4. **EXTERNAL INTERCOSTAL MUSCLES** _____ Orange
5. **INTERNAL INTERCOSTAL MUSCLES** _____ Pink
6. **DIAPHRAGM** _____ Red

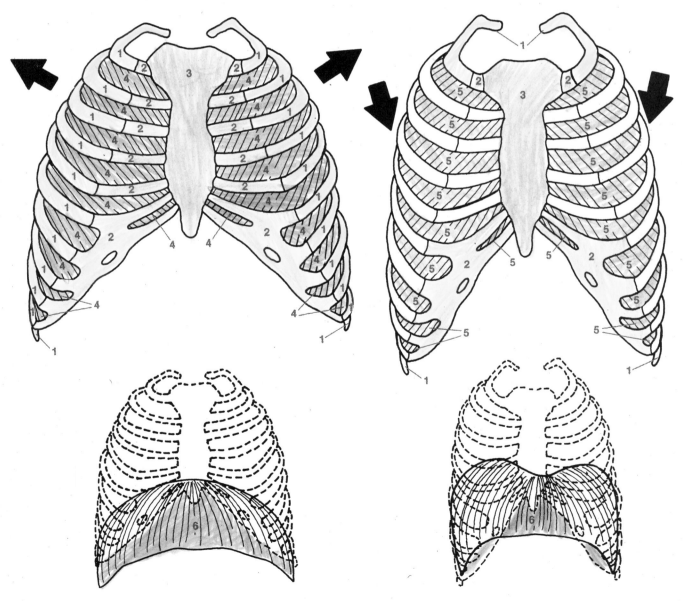

NERVOUS SYSTEM

The nervous system is a control and communication system, consisting of the *brain, spinal cord, nerve cells,* and *nerve fibers,* that runs throughout the body. It originates and coordinates physical reactions to the environment and controls involuntary muscles and organs such as the heart and lungs. It also maintains *homeostasis,* that is, a balanced state within the body.

CENTRAL NERVOUS SYSTEM (CNS)

The brain and spinal cord make up the CNS, the control center for the movement and actions of the entire body. Messages from outlying receptors and sensors arrive at the CNS, where they are interpreted; the CNS then sends out reaction impulses.

PERIPHERAL NERVOUS SYSTEM (PNS)

The first of the two parts of the PNS is the *afferent system,* which carries messages from the sensors to the CNS for processing. The second part, the *efferent system,* carries the CNS's commands to the muscles and organs.

AUTONOMIC NERVOUS SYSTEM (ANS)

The ANS regulates the involuntary internal organs, muscles, and glands.

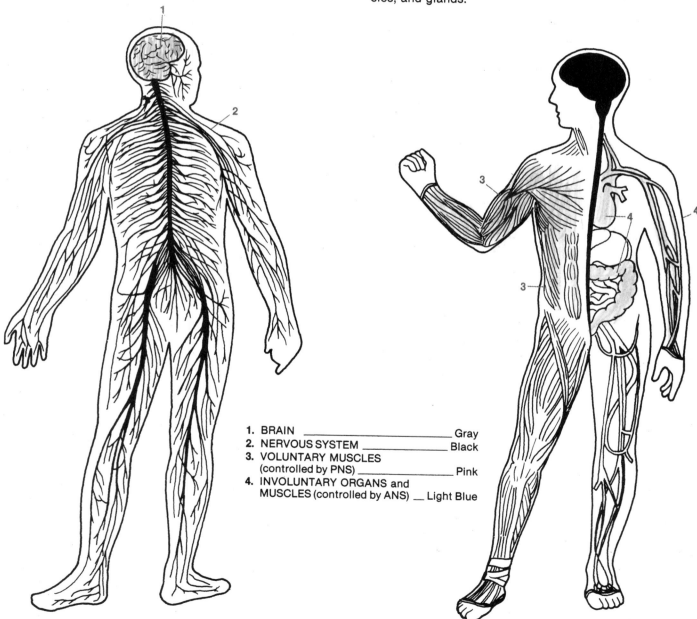

1. BRAIN _____ Gray
2. NERVOUS SYSTEM _____ Black
3. VOLUNTARY MUSCLES
   (controlled by PNS) _____ Pink
4. INVOLUNTARY ORGANS and
   MUSCLES (controlled by ANS) __ Light Blue

NERVE CELL

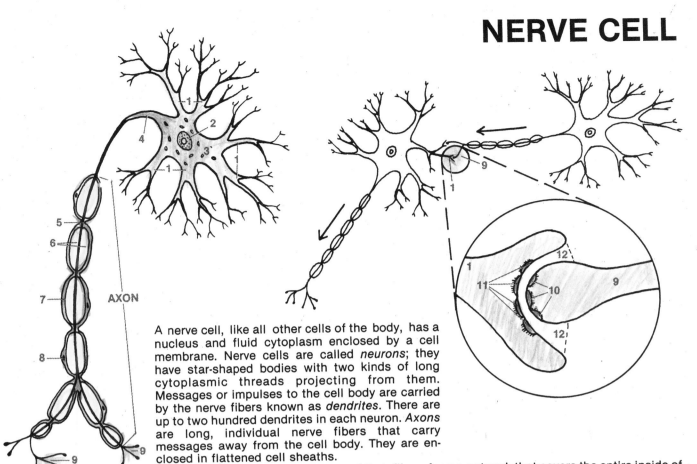

A nerve cell, like all other cells of the body, has a nucleus and fluid cytoplasm enclosed by a cell membrane. Nerve cells are called *neurons*; they have star-shaped bodies with two kinds of long cytoplasmic threads projecting from them. Messages or impulses to the cell body are carried by the nerve fibers known as *dendrites*. There are up to two hundred dendrites in each neuron. *Axons* are long, individual nerve fibers that carry messages away from the cell body. They are enclosed in flattened cell sheaths.

The neurons and their fibers form a network that covers the entire inside of the body and all of the skin. The dendrites receive stimuli from receptor organs or other nerve fibers; an axon then carries the message from the cell body to another neuron or to the organ that is to be effected (called the *effector*) such as a muscle. Neurons that carry messages from sensory organs to the CNS are *sensory neurons*; those that carry messages from the CNS or other nerve centers to muscles or organs are *motor neurons*. The long fibers of neurons are arranged in bundles called *nerves*. The fibers of one neuron never touch those of another, although they do meet at a place called a *synapse*. The nerve impulse is transferred electrochemically through the synapse in an instant.

1. DENDRITE _____ Blue
2. NUCLEUS _____ Green
3. CELL BODY _____ Light Blue
4. AXIS CYLINDER _____ Dark Blue
5. NODE OF RANVIER_____ Orange
6. MYELIN SHEATH _____ Yellow
7. SHEATH OF SCHWANN'S CELL __ Red
8. NUCLEUS OF SCHWANN'S CELL Black
9. TERMINAL BRANCHES _____ Pink
10. PRESYNAPTIC MEMBRANE _____ Gray
11. POSTSYNAPTIC MEMBRANE __ Brown
12. SYNAPTIC CLEFT _____ Light Green

REFLEX ACTION

It takes only a brief moment for an impulse to reach the brain, be evaluated, and return as a muscle command to the appropriate part of the body, but even so that may be too long in times of danger. The reflex action is a safety shortcut. A pain impulse from touching a hot pot or a sharp object, for example, can go from your finger to your spinal cord, where it is transferred to an association neuron that connects directly to the motor neuron. A stimulus is instantly dispatched to jerk your arm away. Had the message travelled to the brain first, there may have been more serious damage to your finger. Blinking, sneezing, coughing, and ducking are all reflex actions.

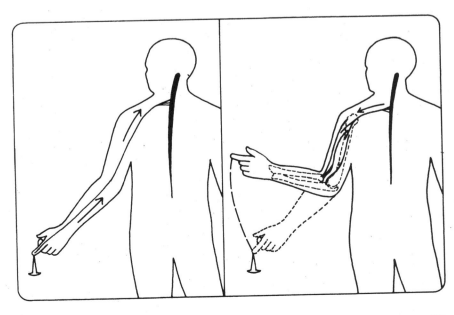

BRANE

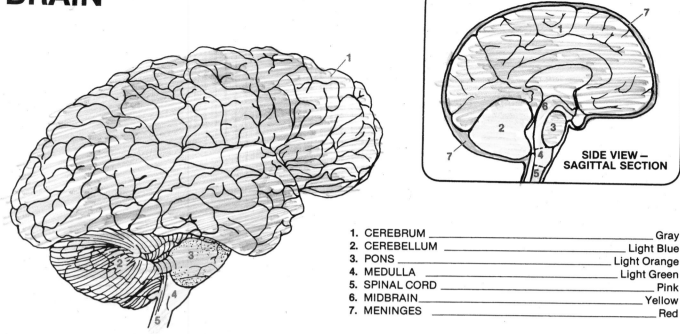

SIDE VIEW – SAGITTAL SECTION

1. CEREBRUM _____ Gray
2. CEREBELLUM _____ Light Blue
3. PONS _____ Light Orange
4. MEDULLA _____ Light Green
5. SPINAL CORD _____ Pink
6. MIDBRAIN _____ Yellow
7. MENINGES _____ Red

The brain is the most complex and specialized organ of the body. It is covered by three protective membranes, the *meninges,* that also extend downward to encase the spinal cord. The main part of the brain, the *cerebrum,* is divided into right and left halves, the *hemispheres.* The outer surface of the cerebrum is the *cortex.* It is wrinkled and irregularly shaped, with deep furrows that increase the brain's surface area. The *forebrain,* located in the front of the cerebrum, is the site of the most complex functions of human thought and action. These functions include memory, judgment, reasoning, speech, and the formation of words. The forebrain also is the seat of emotions and what we know as personality traits, and receives and sends messages to the other parts of the brain that control less complex functions. The *midbrain* controls vision and eye reflexes, many visceral or involuntary muscle activities, and motor responses of the head and torso. The midbrain connects the forebrain to the *hindbrain,* which consists of the *cerebellum* and *pons* and is located behind and below the cerebrum. The hindbrain is responsible for coordinating muscular activity and amplifying cerebral stimuli on their way to the muscles. It cannot initiate a muscular contraction, but it can keep muscles in a state of partial contraction. The pons is a pathway between the two halves of the cerebellum and a relay between the midbrain and the *medulla.* Within the pons is the *pneumotaxic center,* which plays a role in breathing. The *medulla oblongata,* an elongation of the base of the brain that joins with the spinal cord, controls the activity of internal organs, including respiratory and digestive organs, the heart, and glands.

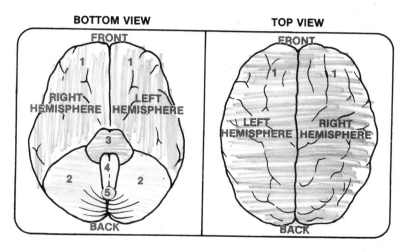

BOTTOM VIEW

FRONT

RIGHT HEMISPHERE   LEFT HEMISPHERE

BACK

TOP VIEW

FRONT

LEFT HEMISPHERE   RIGHT HEMISPHERE

BACK

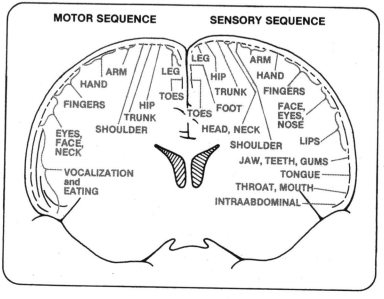

MOTOR SEQUENCE

ARM
HAND
FINGERS
EYES, FACE, NECK
SHOULDER
HIP TRUNK
LEG
TOES
VOCALIZATION and EATING

SENSORY SEQUENCE

LEG
HIP
TRUNK
FOOT
TOES
HEAD, NECK
SHOULDER
ARM
HAND
FINGERS
FACE, EYES, NOSE
LIPS
JAW, TEETH, GUMS
TONGUE
THROAT, MOUTH
INTRAABDOMINAL

SPINAL CORD

The spinal cord descends from the medulla oblongata into the protective armored canal formed by the vertebrae. The center region of the cord is white matter made up of numerous sheathed nerve fibers. The interior is gray matter. The cord has two functions: it serves as the sensory-motor mechanism for reflex actions, and as the two-way transmitter of impulses, reactions, and stimuli triggered by various internal and external conditions. Thirty-one pairs of spinal nerves, the *peripheral nervous system*, branch out from the vertebrae to the right and left sides of the body. These nerves are large cables of sensory and motor fibers. The first nerve pair arises from the medulla, the others from the cord. Each nerve pair controls a particular area of the body and is identified by the number of the vertebra over which it leaves the spinal cord. Outside the cord each nerve divides to form several branches or *rami*, which control general areas of the body. The *dorsal rami* control the muscles and the skin of the back; the *ventral rami* innervate all the structures of the limbs and torso; the *meningeal* or *recurrent* branches return to the spine and vertebrae. The ventral rami and adjacent nerves form networks or braids called *plexuses* that go to general areas. The *cervical plexus* serves the neck, upper shoulders, and the diaphragm via the phrenic nerve; the *brachial plexus* goes to the upper limbs and the neck and shoulder muscles; the *lumbar plexus* controls the abdominal area and part of the legs; and the *sacral plexus* serves the buttocks area and lower legs. Each plexus is like a large cable of nerve bundles—it goes directly to a certain part of the body and specific nerves branch out when the plexus passes the muscle, organ, or tissue it controls.

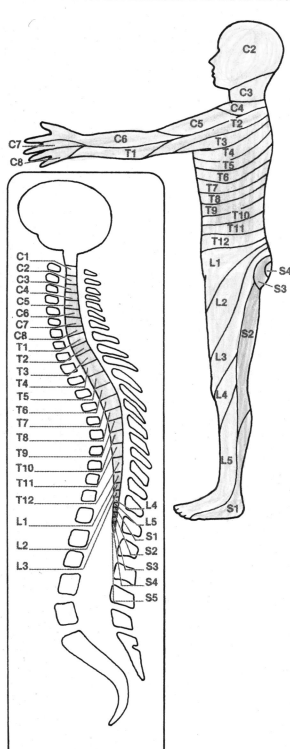

1. VERTEBRA	Light Brown
2. SPINAL CORD	Yellow
3. SPINAL GANGLION	Orange
C1–C8. CERVICAL NERVES	Light Blue
T1–T12. THORACIC NERVES	Purple
L1–L5. LUMBAR NERVES	Pink
S1–S5. SACRAL NERVES	Red

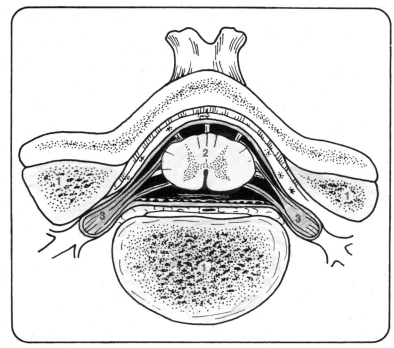

SENSATION SITES

You keep in touch with the outside world with your sensory organs and their nerves and receptors on the surface of your body.

1. SIGHT

Light enters the eye through the lens and is focused on the *retina,* which is lined with photoreceptor neurons called *rods* and *cones.* Rods respond to dim light; cones are stimulated by bright light and are specialized to detect color. The photoreceptors send impulses to *ganglia* (a group of nerve cell bodies) near the front of the retina. The ganglia lead to the *optic nerve,* which in turn transmits impulses to the visual center in the occipital lobe of the brain.

2. SENSATIONS OF THE SKIN

Touch sensors are closest to the skin's surface on your fingertips and near strands of hair. Pressure sensors lie deeper. Pain sensors are bare dendrites. Heat and cold sensors are different from one another and appear randomly over the entire body.

3. SMELL

The *olfactory nerve cells* detect particles given off by objects. The particles generate an impulse that travels over the olfactory tract to the cortex of the brain. After con-stant exposure to an odor, our olfactory nerves become temporarily deadened.

4. TASTE

Food mixes with saliva and enters pores, the *lingual papillae,* on the tongue. Embedded beneath these pores are the chemical receptors that distinguish sweet, salty, bitter, and sour.

5. HEARING

Sound waves enter the *auditory canal* and vibrate the *tympanic membrane* or eardrum. Three tiny bones—the *hammer, anvil,* and *stirrup*—link the inside of the eardrum to the *cochlea,* which is filled with fluid and lined with nerve endings. The vibrations move through the liquid to the sensors and are transmitted to the brain as sound impulses. The *semicircular canals* help to establish our sense of balance.

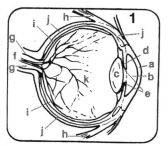

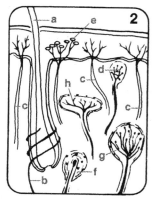

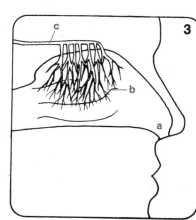

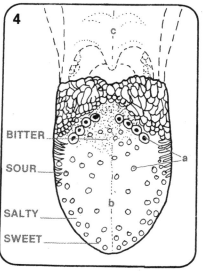

BITTER

SOUR

SALTY

SWEET

1. SIGHT
  a. CORNEA _____ Yellow
  b. ANTERIOR
    CHAMBER _____ Light Brown
  c. LENS _____ Pink
  d. IRIS _____ Blue
  e. PUPIL _____ Gray
  f. RETINAL BLOOD
    VESSELS _____ Red
  g. OPTIC NERVE _____ Green
  h. RECTUS MUSCLE _____ Purple
  i. RETINA _____ Light Green
  j. SCLERA (white of eye) ____ White
  k. POSTERIOR
    CAVITY _____ Light Orange

2. SENSATIONS OF THE SKIN
  a. HAIR _____ Brown
  b. ROOT OF HAIR
    PLEXUS (touch) _____ Black
  c. PAIN RECEPTOR _____ Green
  d. MEISSNER'S
    CORPUSCLE (touch) _____ Pink
  e. MERKEL'S DISCS
    (touch) _____ Orange
  f. PACINIAN CORPUSCLE
    (pressure) _____ Light Blue
  g. END BULB OF KRAUSE
    (cold) _____ Yellow
  h. END ORGAN OF RUFFINI
    (heat) _____ Gray

3. SMELL
  a. NASAL CAVITY _____ Pink

  b. OLFACTORY
    NERVE FIBERS_____ Green
  c. OLFACTORY
    NERVE TRACT_____ Yellow

4. TASTE
  a. PAPILLAE _____ Orange
  b. BODY OF TONGUE _____ Pink
  c. PALATE_____ Gray

5. HEARING
  a. EXTERNAL AUDITORY
    CANAL_____ Flesh
  b. TYMPANIC MEMBRANE
    (eardrum) _____ Red
  c. EUSTACHIAN TUBE ____ Orange
  d. AUDITORY
    BONES _____ Light Brown
  e. SEMICIRCULAR
    CANALS _____ Yellow
  f. COCHLEA _____ Blue
  g. AUDITORY NERVES _____ Green

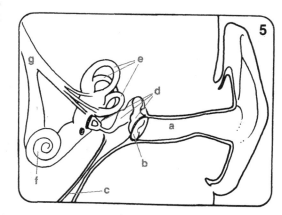

AUTONOMIC NERVOUS SYSTEM

You can blink or move your finger at different speeds, but you can't change how fast your stomach digests food. The stomach, like other internal organs and the smooth muscle tissue of blood vessels, functions involuntarily and is under the control of the autonomic nervous system (ANS). This system regulates the body's life-sustaining functions almost independently of the central nervous system (CNS), which consists of the brain and spinal cord. The ANS is also the division of the peripheral nervous system (PNS) that conveys messages from the CNS outward to produce responses in involuntary muscles and tissues.

There are two main parts to the system: the *sympathetic* and *parasympathetic* systems. Many organs have nerves coming from both systems; these nerves produce opposite reactions. For example, one nerve may respond to low blood pressure by increasing the heartbeat; the other nerve will slow the heartbeat when the blood pressure is too high. Similarly, when the level of carbon dioxide in the blood increases, the cerebral spinal fluid becomes more acidic. The acidity activates chemoreceptors that stimulate the respiratory center of the medulla, which increases the rate of breathing. Increased breathing in turn eliminates excess carbon dioxide, thereby bringing the level in the blood back to normal. If too much carbon dioxide is eliminated, however, the chemoreceptors initiate slower breathing to increase the level of carbon dioxide.

1. GLANDS _____ Yellow-Green
2. EYE _____ Light Brown
3. NASAL MUCOSA _____ Flesh
4. TRACHEA _____ Blue
5. LUNGS _____ Gray
6. HEART _____ Purple
7. LIVER _____ Brown
8. GALLBLADDER _____ Orange
9. STOMACH _____ Green
10. PANCREAS _____ Turquoise
11. a. DUODENUM and b. SMALL INTESTINE _____ Light Blue
12. LARGE INTESTINE _____ Dark Blue
13. RECTUM _____ Light Orange
14. a. FEMALE and b. MALE REPRODUCTIVE ORGANS _____ Yellow
15. KIDNEYS _____ Light Purple
16. URETERS _____ Light Green
17. BLADDER _____ Pink

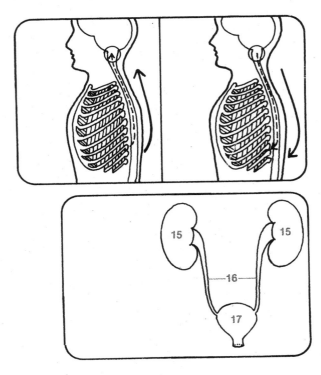

DIGESTIVE SYSTEM

The body cannot use food in the form it ordinarily comes in because the pieces are too large and some foods, such as fats, are not water soluble and therefore cannot be absorbed into the bloodstream or pass into the tissue cells. The chemical complexity of most foods is also more than the body can handle. The role of the digestive system is to reduce large and complex foods to the water-soluble substances the cells can use. The process is both physical — as when the teeth chew meat into tiny bits — and chemical — as when the enzyme ptyalin helps to change starches into smaller compounds. The timing of the digestive system is very important: food must move slowly enough so that all the necessary changes can occur and absorption can take place, but fast enough to prevent harmful decomposition.

There are two parts to the digestive system. The *alimentary canal* is a tube about nine meters long running from the *mouth* to the *anus* and includes the *throat, esophagus, stomach,* and the *small* and *large intestines.* The organs and the glands that aid in the digestive process are the *accessory organs:* the *teeth, tongue, salivary glands, pancreas, liver,* and *gallbladder.* They reduce food mechanically and chemically to a simple form the body can process.

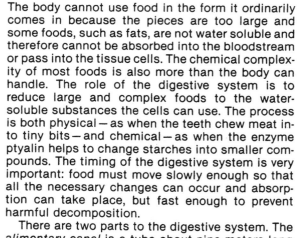

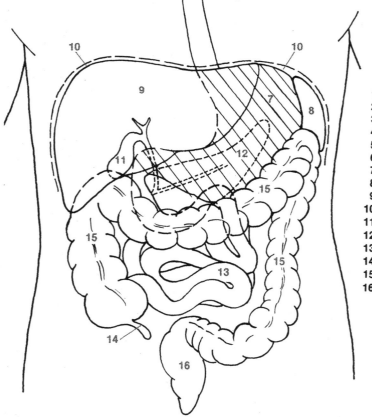

1. TEETH _____ Yellow
2. SALIVARY GLANDS _____ Turquoise
3. TONGUE _____ Pink
4. EPIGLOTTIS _____ Red
5. ESOPHAGUS _____ Yellow-Green
6. TRACHEA _____ Blue
7. STOMACH _____ Green
8. SPLEEN _____ Purple
9. LIVER _____ Brown
10. DIAPHRAGM _____ Flesh
11. GALLBLADDER _____ Orange
12. PANCREAS _____ Light Green
13. SMALL INTESTINE _____ Light Blue
14. APPENDIX _____ Gray
15. LARGE INTESTINE _____ Dark Blue
16. RECTUM _____ Light Orange

MOUTH AND ESOPHAGUS

The mouth's primary role is to help digest food by reducing its size. The mouth includes the *tongue, teeth,* and the *hard* and *soft palates.* There also are three sets of glands that secrete saliva to aid in chewing food: the *parotid glands* are located in front of the ears; the *sublingual glands* are embedded under the sides of the tongue; and the *submaxillary glands* are positioned near the rear of the jawbone. The teeth cut and crush food. The tongue acts as a taste organ and it also mixes saliva with food and moves it toward the rear of the *pharynx* or throat. As food enters the pharynx, the *epiglottis,* a cartilaginous lid that hangs over the larynx to prevent food from entering it, closes and the wave-like swallowing movement begins in the esophagus. The esophagus is a food tube twenty-five to thirty centimeters long that begins at the pharynx and descends through the mediastinum and diaphragm into the stomach. A circular sphincter muscle, located where the esophagus joins the stomach, opens to let food pass and closes behind it to prevent it from flowing back from the stomach.

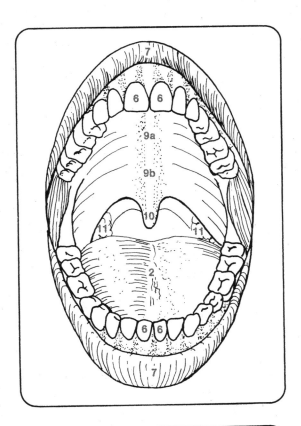

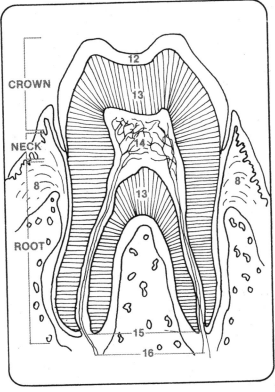

1.	SALIVARY GLANDS	Turquoise
2.	TONGUE	Pink
3.	EPIGLOTTIS	Red
4.	ESOPHAGUS	Yellow-Green
5.	TRACHEA	Blue
6.	TEETH	White
7.	LIPS	Flesh
8.	GINGIVA (gums)	Orange
9.	**a.** HARD and **b.** SOFT PALATES	Light Blue
		Dark Blue
10.	UVULA	Dark Green
11.	TONSIL	White
12.	ENAMEL	Green
13.	DENTIN	Light Brown
14.	PULP CAVITY	Light Orange
15.	ROOT CANAL	Brown
16.	NERVE FIBERS	

STOMACH

The stomach continues the process that began in the mouth of reducing the size of food, but its primary role is storage. The stomach's walls have three layers of smooth muscle— *longitudinal, circular,* and *oblique*—and contractions of these muscles create a twisting, kneading action that breaks up food. The stomach lining is a thick convoluted membrane with many gastric glands embedded in its folds, which form tiny tubes. There are three kinds of glands: *zymogenic* or *chief cells* secrete the enzyme pepsinogen; *mucous cells* secrete mucus and intrinsic factor, which is vital to the absorption of vitamin B_{12}; and *parietal cells* secrete hydrochloric acid, which activates the enzyme and mucous cells. These various secretions begin to digest proteins; during the process, the mucus protects the stomach wall from the acid and pepsinogen. Food moving through the stomach toward the *pyloric antrum* is progressively reduced to finer particles. When the food reaches the pyloric region, a slight pressure builds up that helps meter it through the sphincter valve into the small intestine.

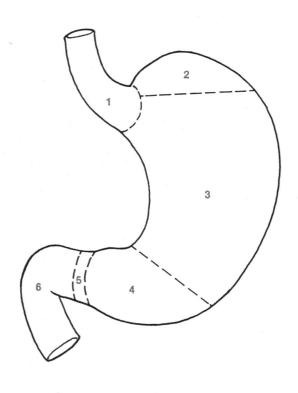

1. ESOPHAGUS _____ Yellow-Green
2. FUNDUS _____ Light Green
3. BODY _____ Green
4. PYLORUS _____ Light Blue
5. PYLORIC SPHINCTER _____ Blue
6. DUODENUM _____ Dark Blue
7. GASTRIC GLANDS _____ Yellow
8. RUGAE _____ Light Orange
9. LAMINA PROPRIA _____ Orange
10. LYMPH NODULE _____ Brown
11. SUBMUCOSA _____ Pink
12. BLOOD VESSELS _____ Red
13. LAYERS OF SMOOTH MUSCLES _____ Flesh
14. **a.** SURFACE and **b.** MUCUS CELLS _____ Light Brown
15. CHIEF CELL _____ Light Purple
16. PARIETAL CELL _____ Gray

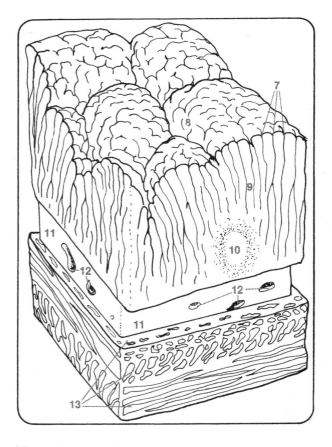

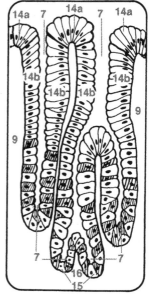

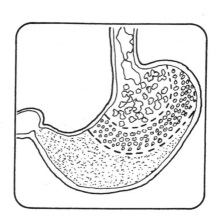

SMALL INTESTINE

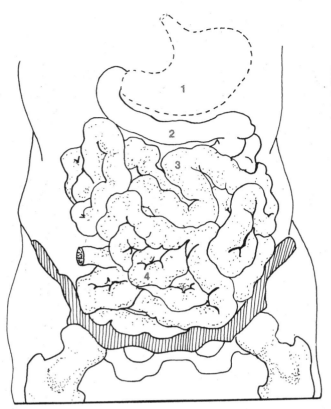

Partially broken-down food leaves the stomach and enters the small intestine, a coiled tube about seven meters long that has three very similar sections: the *duodenum, jejunum,* and, the longest, the *ileum.* The four layers or coats of the intestinal walls are very much like those of the stomach. The walls of the duodenum and of most of the jejunum are arranged in circular folds that increase the surface area of the intestines. This increased surface area makes a greater number of glands available to secrete digestive juices and it also enlarges the area for the absorption of soluble nutrients. The glands, tube-like folds of cells that open onto the surface, secrete enzymes that combine with others from the liver and pancreas. Together these enzymes break food down completely into the simple sugars, fatty acids, triglycerides, and amino acids that can be absorbed into the blood and lymph for distribution to the body's tissues. The intestines are lined with four to five million tiny finger-like projections called *villi.* These move and sway with the intestine's pulsations to mix the food and also to provide a structure for bringing the blood and lymphatic vessels close to the surface. Glycerol and fatty acids enter the villi and are carried away by the lymph. Sugars and amino acids are absorbed into the blood and carried to the liver.

1. STOMACH _____ Green
2. DUODENUM _____ Dark Blue
3. JEJUNUM _____ Light Blue
4. ILEUM _____ Blue
5. **a.** CIRCULAR FOLD
   with **b.** VILLI _____ Light Green
6. MUCOSAL MUSCLE _____ Orange
7. SUBMUCOSA _____ Light Brown
8. CIRCULAR MUSCLE_____ Pink
9. LONGITUDINAL MUSCLE__ Light Purple
10. SEROSA _____ Gray
11. ABSORPTIVE CELLS _____ Turquoise
12. GOBLET CELLS (mucus-
    secreting) _____ Yellow-Green
13. INTESTINAL GLANDS _____ Yellow
14. BLOOD VESSELS _____ Red
15. LYMPHATIC VESSEL _____ Brown

| 25 CENTIMETERS | 2 METERS | 3.7 METERS |

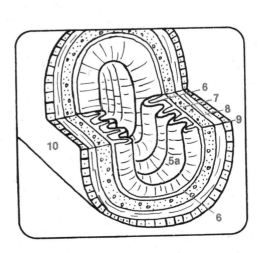

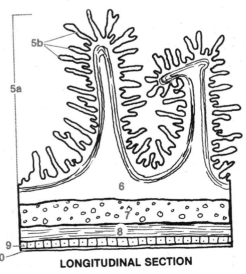

LONGITUDINAL SECTION

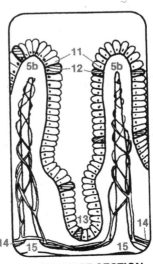

TRANSVERSE SECTION

LARGE INTESTINE

The *large intestine* is 1.5–1.8 meters long and about 7.5 centimeters in diameter. The walls of the large and the small intestines are similar except that the large intestine's have no villi and absorb no nutrients. The most important function of the large intestine is the reabsorption of water and essential electrolytes or salts.

The large and small intestines join in the lower right quadrant of the abdomen. Just below this juncture is the *cecum*, which has a finger-like extension, the *vermiform appendix*, that can become inflamed, a condition known as *appendicitis*. The opening between the ileum and the large intestine is formed by the *ileocecal valve*, two pro-jecting lips that prevent material from flowing back into the small intestine. The large intestine is sometimes called the *colon*, though that term is properly reserved for the portion of the intestine between the cecum and the *rectum*, a muscular tube at the intestine's end. The colon is shaped like an inverted U with an S-curve at its lower end and is divided into four sections: the *ascending colon* runs up the right side of the abdominal cavity; it turns left abruptly to become the *transverse colon*, which runs across the top of the cavity; the transverse colon plunges on the left side to form the *descending colon;* and finally the colon ends with the S-curve, the *sigmoid colon*. The colon receives a watery mass of undigested food within two to five hours after eating; as this mass passes through it, much of the water is absorbed. This mass, called *feces,* becomes more dehydrated and passes into the rectum for elimination. There are internal and external sphincter valves at the end of the rectum that remain tightly closed except during defecation.

If injury or disease prevents the intestine and rectum from being used, the colon can be opened surgically to allow waste removal, an operation called a *colostomy*. For a temporary condition, the abdomen is slit and the colon is brought to the surface, cut open, and drained into a bag-like container. When the problem is over, the colon is sutured together and reembedded in the abdomen. A permanent condition requires that the colon be cut and a single end be brought to the surface to be fitted with an adaptor and bag for waste drainage.

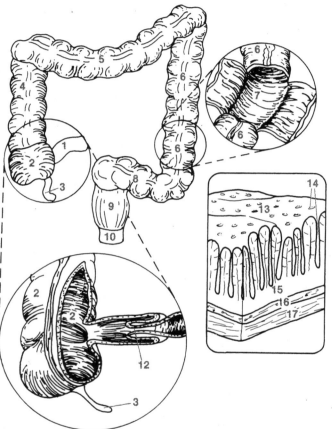

1. ILEUM _____ Blue
2. CECUM _____ Yellow
3. APPENDIX _____ Gray
4. ASCENDING COLON _____ Yellow-Green
5. TRANSVERSE COLON _____ Light Green
6. DESCENDING COLON _____ Green
7. INTERIOR OF DESCENDING COLON _____ Light Blue
8. SIGMOID COLON _____ Turquoise
9. RECTUM _____ Light Orange
10. ANAL CANAL and ANUS _____ Orange
11. RECTAL TRANSVERSE FOLDS _____ Pink
12. ILEOCECAL (colic) VALVE _____ Dark Blue
13. EPITHELIUM _____ Flesh
14. INTESTINAL GLANDS and OPENINGS _____ Brown
15. LAMINA PROPRIA _____ Purple
16. SUBMUCOSA _____ Light Brown
17. SMOOTH MUSCLE _____ Light Purple

DOUBLE-BARREL COLOSTOMY (TEMPORARY)

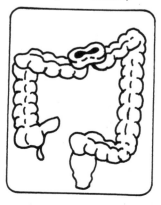

COLOSTOMY (PERMANENT)

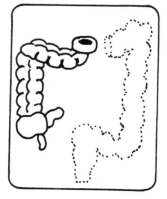

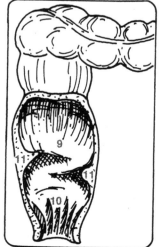

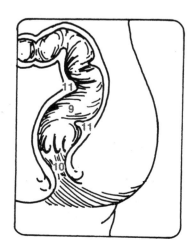

34

ACCESSORY ORGANS OF DIGESTION

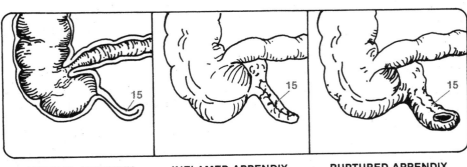

NORMAL APPENDIX   INFLAMED APPENDIX   RUPTURED APPENDIX

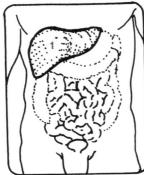

The human *appendix* does not serve as an accessory organ of digestion, but it does play a role in digestion in the animals from which humans evolved. The appendix may become inflamed and even rupture.

The *liver* is the body's largest gland. It weighs 1.26 kilograms and secretes *bile*, a fluid that increases the solubility of fats in water. The *bile ducts* form a large Y-shaped tube; bile travels down one leg of the Y to the duodenum and up the other for storage in the gallbladder. A large amount of venous blood goes to the liver. The liver's most important job is to reduce or remove toxic chemicals from the bloodstream. The *spleen* assists the liver by removing damaged blood cells.

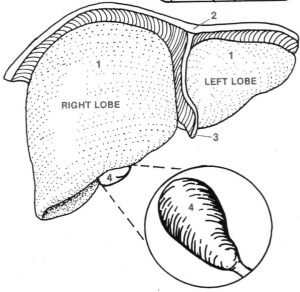

RIGHT LOBE

LEFT LOBE

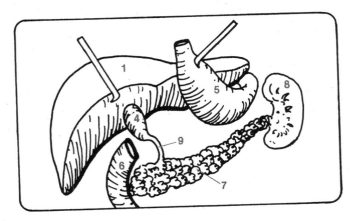

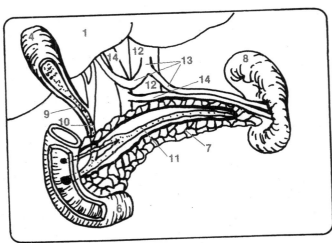

The *pancreas*, located behind the stomach and above the intestine, produces a fluid with three enzymes that breaks down all classes of organic nutrients and unites with bile in the *common bile duct*, from where it is secreted into the intestine. The pancreas also secretes insulin from the special beta cells of the *islets of Langerhans*. Without insulin, sugar collects in the blood instead of reaching the tissues, a condition known as *sugar diabetes*.

The *gallbladder* is a pear-shaped sac on the underside of the liver. It receives most of the liver's bile and stores it until needed, reacting quickly when acidic gastric juices and fatty foods enter the small intestine.

1. LIVER _____ Brown
2. DIAPHRAGM _____ Red
3. FALCIFORM LIGAMENT _____ Flesh
4. GALLBLADDER _____ Orange
5. STOMACH _____ Green
6. DUODENUM_____ Dark Blue
7. PANCREAS _____ Light Green
8. SPLEEN _____ Purple
9. CYSTIC DUCT _____ Light Orange
10. COMMON BILE DUCT_____ Yellow
11. PANCREATIC DUCT _____ Yellow-Green
12. AORTA _____ Light Purple
13. CELIAC, SPLENIC, and HEPATIC ARTERIES _____ Pink
14. SPLENIC and HEPATIC PORTAL VEINS _____ Light Blue
15. APPENDIX _____ Gray

URINARY SYSTEM

The body constantly produces not only solid wastes but also very complex and often toxic chemical compounds. Tissues filled with these wastes cannot absorb food or oxygen. The excretory organs of the urinary system eliminate liquid wastes. Each of the body's cells discharges its wastes into the bloodstream. The blood carries the acids and salts to the bean-shaped *kidneys* located on either side of the spine in the lumbar area. Each kidney has about 1,250,000 *nephrons* or filters that control the chemical composition of the blood. The nephrons filter the blood; remove nitrogen-compound wastes, excess water, and salts; and return necessary substances and fluids to the blood. The waste-filled fluids migrate through small nephron tubules to larger tubules in the kidney and finally into the *ureter*, a long tube that descends into the *urinary bladder*. The bladder is a hollow muscular organ. When it fills with the slightly acidic yellow fluid called *urine* (about 1,200–1,500 milliliters are produced every day), a reflex action relaxes the inner sphincter muscle in the lower bladder. *Micturition*, the act of expelling urine, is voluntarily started by relaxing the external sphincter. In the female the urethra empties in the area between the clitoris and the vaginal opening. In the male the urethra is about twenty centimeters long and runs through the penis.

1. RIBS _____ Yellow
2. KIDNEY _____ Light Purple
3. URETER _____ Light Blue
4. BLADDER _____ Pink
5. URETHRA _____ Purple
6. RECTUM _____ Light Orange
7. ANUS _____ Orange

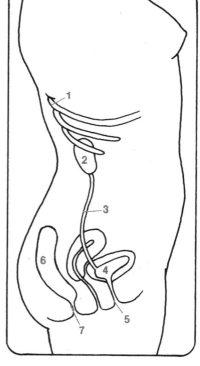

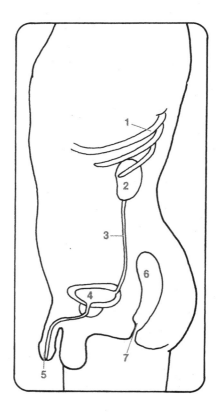

KIDNEYS

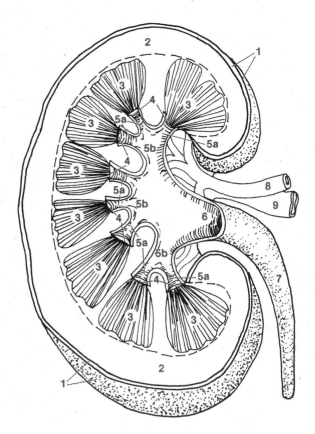

The two kidneys filter out chemicals and electrolytes from the bloodstream. Blood enters each kidney from the large *renal artery* that comes directly from the aorta. The artery divides and subdivides into a maze of arterioles, and each arteriole ends in a coil of capillaries, a *glomerulus.* The coils insert into a cup-like part of the nephron, *Bowman's capsule.* A great deal of water, wastes, glucose, and salts are filtered into each capsule. From the capsule the fluid enters a tubule that passes through a network of capillaries. Many of the fluids, compounds, and minerals diffuse back into the blood and return to the bloodstream. Nitrogenous wastes, excess water, and salts pass into increasingly larger tubules and flow into the *renal pelvis* as urine. Blood returning from the kidney to the bloodstream has very few impurities in it.

The kidneys have a great deal of extra capability. If one kidney is destroyed or removed, the other becomes enlarged and can provide the filtration of the original two. The kidneys also help to maintain the body's delicate acid-alkaline balance by excreting or reabsorbing acidic hydrogen or alkaline bicarbonate ions.

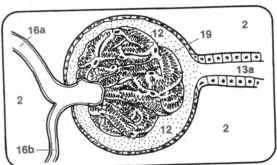

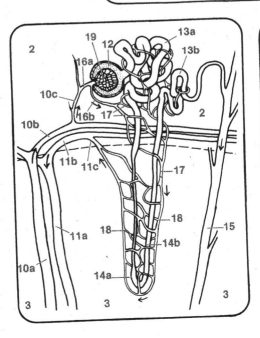

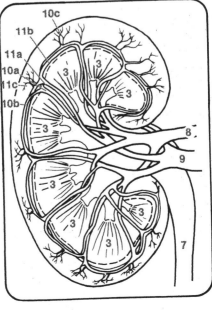

1. RENAL CAPSULE _____ Light Purple
2. CORTEX _____ Flesh
3. MEDULLA (pyramid) _ Orange
4. PAPILLA _____ Yellow
5. a. MINOR and
   b. MAJOR CALYCES _ Green
6. RENAL PELVIS _ Light Green
7. URETER _____ Light Blue
8. RENAL ARTERY _____ Red
9. RENAL VEIN _____ Dark Blue
10. a. INTERLOBAR,
    b. ARCUATE, and
    c. INTERLOBULAR ARTERIES _____ Red
11. a. INTERLOBAR,
    b. ARCUATE, and
    c. INTERLOBULAR VEINS _____ Blue
12. BOWMAN'S CAPSULE _ Gray
13. a. PROXIMAL and
    b. DISTAL CONVOLUTED TUBES _____ Light Brown
14. a. DESCENDING and
    b. ASCENDING HENLE'S LOOPS _____ Brown
15. COLLECTING DUCT _____ Light Orange
16. a. AFFERENT and
    b. EFFERENT ARTERIOLES _____ Red
17. ARTERIOLE _____ Red
18. VENULE RECTAE _____ Blue
19. GLOMERULAR CAPILLARIES _____ Purple

REPRODUCTIVE SYSTEM

The organs of the reproductive system create human off-spring by combining genes from the male and female. These organs also help to develop and nurture the genetically unique fetus. The male and female reproductive organs differ in function and appearance. During sexual intercourse the male provides the sperm cell that fertilizes the female's egg, from which the fetus will develop.

MALE REPRODUCTIVE SYSTEM

Sperm are produced in the two *testes* or male gonads, which are contained in the *scrotum*, a pouch suspended outside the body. The sperm are stored in the *epididymis* and pass through the *vas deferens*, ducts that lead to the *seminal vesicles* on each side of the bladder. During the sexual act the *penis* enlarges and becomes rigid or *erect* and the sperm passes through the *prostate gland*, which secretes a milky fluid that enhances the sperm's movement, and *Cowper's gland*, which secretes a protective and lubricating mucus. The sperm and these fluids together are called *semen*. The semen is ejaculated into the *urethra* and penis and, from there, into the female's vagina. Although only one sperm cell is needed to fertilize the egg, there are millions in each ejaculation, which greatly increases the possibility that fertilization will occur during a particular instance of intercourse.

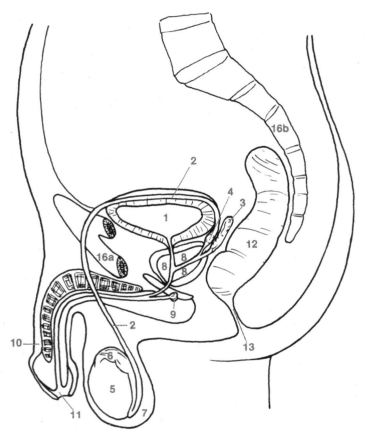

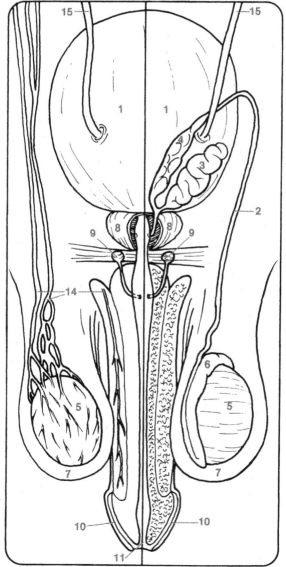

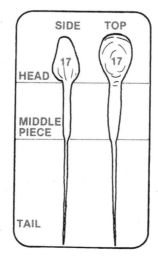

1. BLADDER _____ Pink
2. VAS DEFERENS _____ Yellow-Green
3. SEMINAL VESICLE _____ Yellow
4. AMPULLA OF VAS DEFERENS __ Green
5. TESTIS _____ Light Green
6. HEAD OF EPIDIDYMIS _____ Turquoise
7. SCROTUM _____ Light Purple
8. PROSTATE GLAND _____ Brown
9. COWPER'S GLAND _____ Blue
10. PENIS_____ Flesh
11. URETHRA _____ Purple
12. RECTUM_____ Light Orange
13. ANUS _____ Orange
14. BLOOD VESSELS _____ Red
15. URETER from KIDNEY _____ Light Blue
16. **a.**PUBIS BONE and
    b. SPINE _____ Light Brown
17. SPERM _____ Gray

FEMALE REPRODUCTIVE SYSTEM

The *vagina*, a tube-like potential space seven to ten centimeters long, receives the sperm from the male. The group of external organs surrounding the vagina are the *vulva*. The sperm must reach the *uterus* or womb, a hollow pear-shaped organ, through the *cervix*, the narrow end of the uterus. The uterus is lined with a mucous membrane and has small glands and many capillaries. At a certain time during the twenty-eight-day *menstrual cycle*, a *follicle* in the ovaries produces an *ovum* or egg, which is then carried into the *fallopian tubes* by ciliated cells in the tubes' lining. The cilia also assist the sperm as they swim up the tubes toward the egg. If a sperm enters the egg, a membrane that keeps out other sperm forms around the fertilized egg, now called a *zygote*. The zygote passes into the uterus in three to five days and becomes attached to its lining. The cells begin to multiply and develop into a fetus. During the nine months of gestation, the fetus is enclosed in the *placenta*, a membranous sac, and is nourished by and receives oxygen from the mother's blood passing through the membrane via the *umbilical cord*. Wastes return through the placenta to the mother and are eliminated through her respiratory and excretory systems. As the fetus is developing, the mother's two *mammary glands* in her breasts begin to enlarge. These glands are composed of adipose as well as glandular tissue and secrete milk after the baby is born.

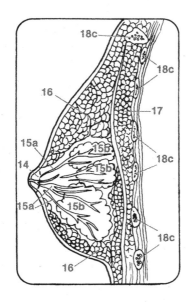

1. OVARY _____ Light Green
2. FALLOPIAN TUBE _____ Yellow-Green
3. UTERUS _____ Green
4. CERVIX _____ Blue
5. VAGINA _____ Light Blue
6. BLADDER _____ Pink
7. URETHRA _____ Purple
8. CLITORIS _____ Brown
9. LABIA _____ Yellow
10. ANUS _____ Orange
11. RECTUM _____ Light Orange
12. FETUS _____ Flesh
13. UMBILICAL CORD _____ Light Purple
14. NIPPLE _____ Red
15. a. MAMMARY DUCTS and
    b. LOBES _____ Turquoise
16. FAT IN SUPERFICIAL FASCIA_____ Gray
17. PECTORALIS MAJOR (muscle) _____ Dark Blue
18. a. PUBIS BONE, b. SPINE, and
    c. RIBS _____ Light Brown

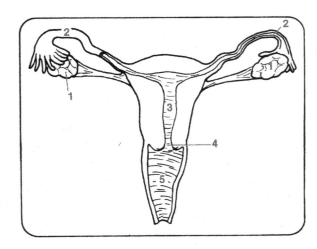

ENDOCRINE SYSTEM

Hormones are essential to our understanding of the body's metabolism or normal functioning. Endocrine glands secrete hormones directly into the bloodstream, not just to a special organ; thus these secretions reach every part of the body. Hormones influence the flow of substances through cell membranes and often work together, which means that a hormonal imbalance may interfere with normal body functions.

THYMUS

The thymus is not always classed as an endocrine gland. It becomes most developed during a child's early years. Apparently its purpose is to initiate antibody formation in the blood.

THYROID

The two lobes of the thyroid gland are located on either side of the trachea and secrete iodine-based hormones that regulate physical and mental growth, oxidation, heart rate, blood pressure, temperature, glucose absorption, and the utilization of glucose.

PARATHYROID

There are four parathyroid glands, all located next to the thyroid. Their secretions control the use of calcium in bone growth, muscle tone, and nervous activity.

PITUITARY

Located at the base of the brain, the pituitary gland secretes hormones that influence other glands. The pituitary gland regulates skeletal growth, the development of the reproductive organs, secretions from the ovaries and testes, the stimulation of the mammary glands to provide milk, blood pressure, the performance of smooth muscles, the reabsorption of water in the kidneys, and the functioning of the adrenal cortex, which becomes more active during times of stress. Pituitary disorders may result in giantism or dwarfism.

PINEAL

The function of the pineal gland is unknown, but it is very active metabolically. It is about the size of a pea and located at the base of the brain.

ADRENAL

Located above the kidneys, the adrenal gland secretes *cortisol,* which regulates the metabolism and the balance between salt and water levels. During emergencies it also secretes *adrenaline (epinephrine),* which increases the heart rate and stimulates the liver and nervous system.

PANCREAS

The level of sugar in the blood is controlled by the pancreas's secretion, *insulin.* Sugar diabetes results when the level of insulin in the blood is relatively low.

OVARIES

Ovaries are found only in women. They secrete the two female hormones—*estrogen,* which produces female characteristics and initiates female bodily functions; and *progesterone,* which affects the endometrial lining of the uterus.

TESTES

Only men have testes. They secrete *testosterone,* the male hormone, which controls the growth of body hair and beard, body size, and the deepening of the voice.

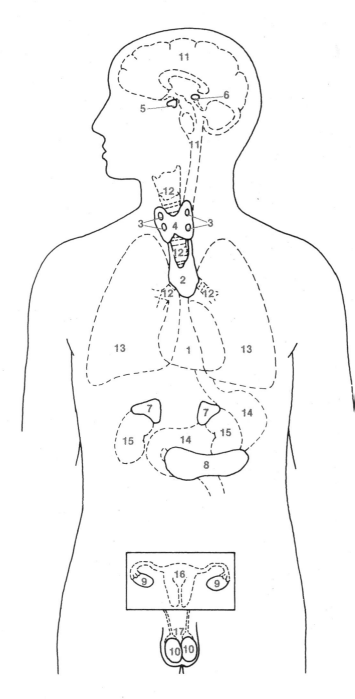

1. HEART _____ Purple
2. THYMUS GLAND _____ Orange
3. PARATHYROID GLAND _____ Pink
4. THYROID_____ Brown
5. PITUITARY GLAND _____ Turquoise
6. PINEAL GLAND _____ Light Brown
7. ADRENAL GLAND_____ Red
8. PANCREAS _____ Light Green
9. OVARIES_____ Yellow
10. TESTES _____ Dark Green
11. BRAIN and SPINAL CORD_____ Gray
12. TRACHEA and BRONCHUS _____ Blue
13. LUNGS _____ Light Blue
14. STOMACH _____ Green
15. KIDNEYS _____ Light Purple
16. UTERUS and FALLOPIAN TUBES _____ Yellow-Green
17. SCROTUM _____ Flesh

LYMPH SYSTEM

Surrounding the cells is a fluid that picks up colloids, particles, electrolytes, and wastes that are unable to return to the blood. The lymph system, a system of tubes like the venous system (their vessels have similar walls and valves to prevent the backflow of fluid), drains the fluid, which is called *lymph* when it enters the system. The microscopic lymph capillaries are larger and more permeable than blood capillaries. Lymph vessels unite to form larger and larger lymphatics. At intervals along the system, the vessels form *lymph nodes* that strain and purify lymph and remove infectious organisms and bacteria before the lymph returns to the blood. The greatest concentration of nodes is in the neck, armpit, elbow, and groin. The *tonsils* and *adenoids* are also masses of lymphatic tissue. Lymphatics from the right side of the head, neck, and right arm flow from the *right lymphatic duct* into the *right subclavian vein*. The lymphatics from the rest of the body drain into the *thoracic duct*, which in turn flows into the *left subclavian vein*.

SPLEEN

The spleen stores large quantities of blood and, in the fetus and newborn child, forms red blood cells. It also removes damaged cells, bacteria, and debris from the blood.

1. EFFERENT LYMPHATIC VESSEL_____ Red
2. AFFERENT LYMPHATIC VESSEL _____ Pink
3. LYMPHATIC VALVE _____ Orange
4. HILUM _____ Light Blue
5. PRIMARY LYMPH NODULE_____ Flesh
6. MEDULLARY CORD _____ Green
7. CORTEX _____ Yellow
8. MEDULLA _____ Blue

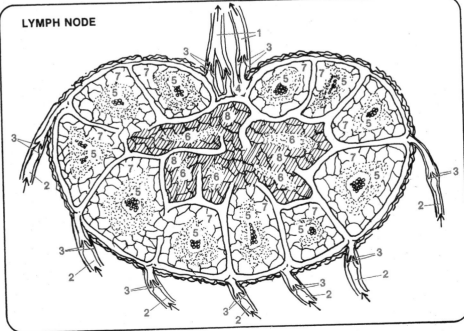

LYMPH NODE

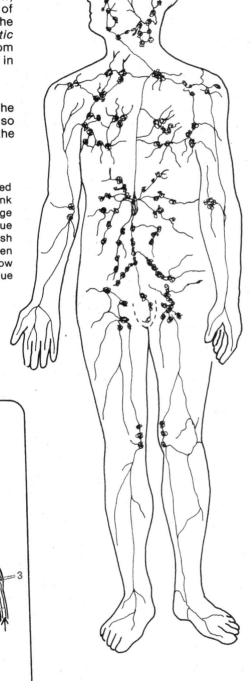

SKIN

The waterproof and airtight skin, the body's largest organ, completely encases the other organs and tissues. Skin protects you from many infectious organisms and harmful light rays, and it also converts light to vitamin D, helps to control the body's temperature, and eliminates certain wastes. Sensors in your skin detect pressure, pain, and the temperature outside your body.

The outer layer of the skin is called the *epidermis;* the thicker inner layer is the *dermis.* The epidermis has a different number of sublayers depending on how much wear it must take. The palms of the hands and soles of the feet have five sublayers; elsewhere there are four. Cells germinate at the innermost layer, the *stratum germinativum.* As the cells multiply, they move upward and undergo chemical changes until they dry and die and flake off. This upward movement and the transformation from moist to dry permits the inner tissues to be surrounded by fluid even though your body exists in the relatively dry air.

The dermis contains numerous blood vessels, nerves, glands, and hair follicles. Fibers from the dermis extend into the subcutaneous layer and anchor the skin.

SEBACEOUS (Oil) GLANDS

The sebaceous glands are located all over the body except on the palms and soles. They are connected to hair follicles by a short duct and secrete *sebum,* an oil that keeps skin soft and hair from drying and becoming brittle. When you are cold or frightened, a muscle (*arrector pili*) attached to the hair follicle pulls the hair erect, which ejects oil onto the skin to prevent evaporation and heat loss. When pushed up, the glands form goose bumps.

SUDORIFEROUS (Sweat) GLANDS

Sweat glands are found over the entire skin, but they are most numerous in the palms, soles, forehead, and armpits. Each gland is a single coiled tube that originates in the subcutaneous tissue, passes through the dermis, and opens as a pore in the epidermis. Its base is surrounded by capillaries from which it extracts the salts, water, and acids from the blood that then are excreted as perspiration.

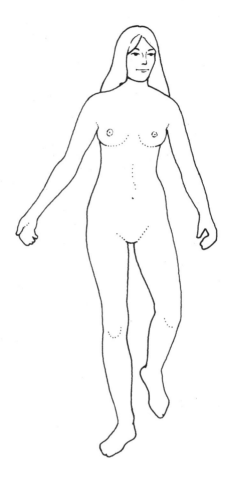

1. LAYER OF OIL _____ Light Green
2. STRATUM CORNEUM _____ Flesh
3. STRATUM LUCIDUM _____ Orange
4. STRATUM GRANULOSUM _____ Pink
5. STRATUM GERMINATIVUM _____ Light Orange
6. SWEAT GLAND DUCT (pore) _____ Yellow-Green
7. SUDORIFEROUS (sweat) GLAND _____ Yellow
8. SEBACEOUS (oil) GLAND _____ Green
9. MUSCLE (arrector pili) _____ Red
10. HAIR _____ Black
11. DERMIS _____ Light Brown
12. HAIR FOLLICLE _____ Gray

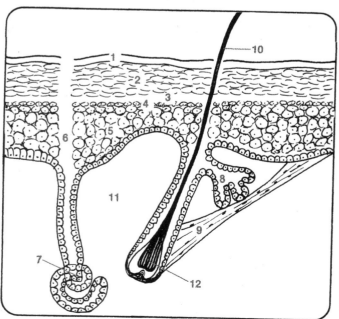

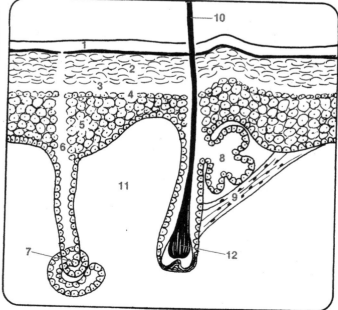

SCARRING, NAILS, AND COOLING

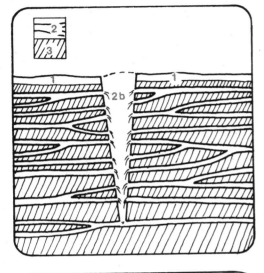

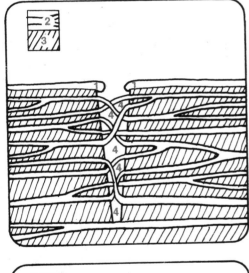

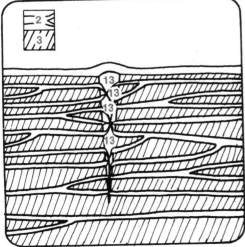

Normal skin tissue does not grow back after a large wound; *scar tissue* forms instead. When a wound occurs, blood rushes in to clean and seal it by clotting and forming a scab. Then the area begins to fill with granular tissue made of blood vessels and newly formed, hard connective tissue. The epidermis bridges the top. After the wounded tissue is replaced, the vessels wither, stopping the supply of blood. The absence of blood makes the scar tissue whitish.

NAILS

Look at your fingers. The base of each nail, the *nail root,* is hidden beneath skin. As a nail grows, it slides forward over the *nail bed,* which looks pink because of the blood in the capillaries just beneath it. A white crescent appears at the base of the nail because the capillaries there are not close to the skin's surface. The visible part of the nail is the *nail body.* Fingernails grow about one millimeter a week, toenails more slowly.

COOLING BY PERSPIRATION

The evaporation of perspiration cools the body because as water turns into water vapor, the vapor draws off excess heat. A similar thing happens to your tea kettle: your skin is like the burner that gives heat to the water until the water changes into a vapor and leaves the surface, taking the heat with it.

1. EPIDERMIS _____ Flesh
2. a. BLOOD VESSELS and
   b. BLOOD _____ Red
3. DERMIS _____ Pink
4. GRANULAR TISSUE_____ Yellow
5. PERSPIRATION _____ Blue
6. HEAT _____Orange
7. a. NAIL BODY and
   b. NAIL ROOT_____ Yellow-Green
8. EPONYCHIUM (cuticle)____ Green
9. NAIL MATRIX _____ Purple
10. STRATUM
    GERMINATIVUM_____ Light Purple
11. STRATUM
    GRANULOSUM ____ Light Brown
12. STRATUM
    CORNEUM _____ Light Orange
13. SCAR TISSUE_____ **White**

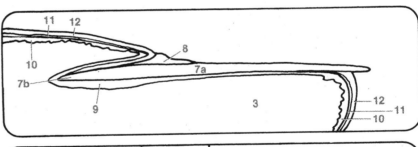

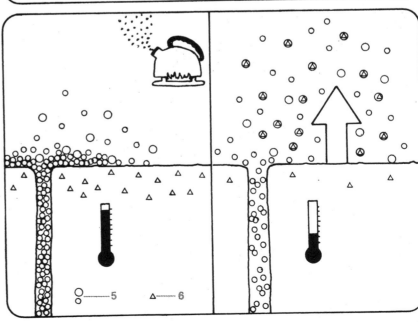